JRN

THE PREVENTION OF CORROSION

By the same author

THE CHEMISTRY OF BUILDING MATERIALS
(APPLIED CHEMISTRY SERIES, NUMBER 1)

THE PREVENTION
OF CORROSION

R. M. E. DIAMANT, MSc, DipChemE, AMInstF
Lecturer in Chemistry and Applied Chemistry
University of Salford

Applied Chemistry Series
Number 2
General Editor
Professor G. R. Ramage, University of Salford

BUSINESS BOOKS LIMITED
London

First published 1971

ISBN 0 220 66889 2

This book has been set 10 on 11 pt Times Roman and
printed in England by Clarke, Doble & Brendon Ltd.
for the publishers, Business Books Limited
(Registered office: 180 Fleet Street, London EC4);
Publishing offices: Mercury House, Waterloo Road, London SE1

MADE AND PRINTED IN GREAT BRITAIN

Contents

List of Figures

Preface

Corrosion is responsible for annual losses amounting to millions of pounds.

The reason for this is certainly not a lack of fundamental research in the field. Ulick R. Evans, FRS, one of the world's foremost authorities on corrosion, lists in his filing system over 9,000 research publications dealing with corrosion.

Despite the amount of scientific research carried out in the field of corrosion over the years, the average engineer seems to know little about the subject. Numerous really serious design errors from the corrosion aspect are constantly being committed, even by the largest industrial undertakings. For example, in the motor industry quite elementary design blunders have been made, causing heavy and often dangerous corrosion to take place in vehicles.

The fault must lie in bad communication. In the science of corrosion prevention, as in other fields, there is a huge chasm between the scientist in his laboratory and the practical designer in his drawing office.

This book is intended to bridge the gap. Its aim is the same as its title: 'The prevention of corrosion'. It is meant as a textbook for engineering and metallurgy students, and as a concise practical guide for engineers in industry. Basic theoretical treatment is confined to the first chapter, the rest being mainly devoted to practical aspects of corrosion and its prevention.

At the end of each chapter there is a bibliography to enable readers to study the subject further. It must be appreciated that this fairly short book cannot hope to do more than to serve as an introduction to this huge subject, but I am convinced that this is precisely what is needed. No attempt is made to provide a comprehensive work and those requiring to know more about the subject are referred to the treatises published by U. R. Evans and L. L. Schreir

EVANS, U. R., *The corrosion and oxidation of metals*, Arnold, London (1960)
——, *The corrosion and oxidation of metals*, First supplementary volume, Arnold, London (1968)
SCHREIR, L. L., *Corrosion*, 2 volumes, Newnes, London (1963)

I would like to express my thanks to Professor G. R. Ramage, Chairman of the Department of Chemistry and Applied Chemistry at the University of Salford, for the help he has given me in the preparation of this book.

R. M. E. DIAMANT
University of Salford

SI Conversion Factors

Length   1 yard=0·914 4 metre
         1 foot=0·304 8 metre
         1 inch=25·4 millimetre

Area   1 yard^2=0·836 127 metre^2
         1 foot^2=0·092 903 metre^2
         1 inch^2=645·16 millimetre^2

Volume   1 yard^3=0·764 555 metre^3
         1 foot^3=28·316 8 decimetre^3 (litre)
         1 inch^3=16·387 1 centimetre^3

Capacity   1 UK gallon=4·546 09 decimetre^3 (litre)
         1 US gallon=3·785 41 decimetre^3 (litre)

Mass   1 UK ton=1,016·05 kilogramme
         1 US ton=907·184 4 kilogramme
         1 hundredweight (UK)=50·802 3 kilogramme
         1 pound=0·453 592 37 kilogramme
         1 ounce=28·349 5 gramme
         1 grain=0·064 798 9 gramme

Temperature $°F=9/5°C+32$
         $K=°C+273·15$
         $°R=1·8 K$

Heat   1 British thermal unit=1·055 06 kilojoule
         1 centigrade heat unit=1·899 108 kilojoule
         1 kilocalorie=4·186 8 kilojoule

Power   1 horsepower=0·745 7 kilowatt
         1 horsepower (metric)=0.735 499 kilowatt

Pressure   1 atmosphere=101·325 kilonewton/metre^2
         1 inch of mercury=3·386 389 kilonewton/metre^2
         1 centimetre of mercury=1·333 22 kilonewton/metre^2
         1 bar=10^5 newton/metre^2
         1 inch water gauge=249·082 newton/metre^2
         1 centimetre water gauge=98·063 8 newton/metre^2
         1 pound (f)/foot^2=1·488 164 newton/metre^2
         1 pound (f)/inch^2=6·894 76 kilonewton/metre^2
         1 ton (f) (UK)/foot^2=107·252 kilonewton/metre^2

> 1 ton (f) (UK)/inch^2 = 15·444 3 meganewton/metre^2 = 15·444 3 newton/millimetre^2

Density
> 1 ton (UK)/yard^3 = 1·328 94 tonne/metre^3
> 1 pound/foot^3 = 16·018 5 kilogramme/metre^3
> 1 pound/inch^3 = 27·679 9 kilogramme/decimetre^3

Concentration
> 1 grain/100 foot^3 = 0·022 883 5 gramme/metre^3
> 1 ounce/gallon (UK) = 6·236 gramme/decimetre^3
> 1 grain/gallon (UK) = 14·254 gramme/metre^3

Thermal conductivity
> 1 Btu/foot hour degF = 1·730 73 watt/metre degC
> 1 Btu inch/foot^2 hour degF = 0·144 228 watt/metre degC
> 1 1 kilocalorie/metre hour degC = 1·163 watt/metre degC

Thermal conductance
> 1 Btu/foot^2 hour degF = 5·678 26 watt/metre^2 degC
> 1 kilocalorie/metre^2 hour degC = 1·163 watt/metre^2 degC

Moisture and air diffusivity

$$1 \text{ grain inch/foot}^2 \text{ inch mercury hour} = 1\cdot453\,\frac{\text{milligramme metre}}{\text{meganewton second}}$$

$$= 5\cdot231\ 6\frac{\text{gramme metre}}{\text{meganewton hour}}$$

Note: Thermal conductivities can be expressed as either watt/metre degC or watt/metre kelvin

BASIC SI UNITS

Physical quantity	Name of unit	Symbol for unit
Length	metre	m
Mass	kilogramme	kg
Time	second	s
Electric current	ampere	A
Thermodynamic temperature	kelvin	K
Luminous intensity	candela	cd

Symbols for units do not take a plural form.

SUPPLEMENTARY UNITS

Physical quantity	Name of unit	Symbol for unit	Definition of unit
Plane angle	radian	rad	} Dimensionless
Solid angle	steradian	sr	
Energy	joule	J	$kg\ m^2\ s^{-2}$
Force	newton	N	$kg\ m\ s^{-2} = J\ m^{-1}$
Power	watt	W	$kg\ m^2\ s^{-3} = J\ s^{-1}$
Electric charge	coulomb	C	$A\ s$
Electric potential difference	volt	V	$kg\ m^2\ s^{-3}\ A^{-1} = J\ A^{-1}\ s^{-1}$
Electric resistance	ohm	Ω	$kg\ m^2\ s^{-3}\ A^{-2} = V\ A^{-1}$
Electric capacitance	farad	F	$A^2\ s^4\ kg^{-1}\ m^{-2} = A\ s\ V^{-1}$
Magnetic flux	weber	Wb	$kg\ m^2\ s^{-2}\ A^{-1} = V\ s$
Inductance	henry	H	$kg\ m^2\ s^{-2}\ A^{-2} = V\ s\ A^{-1}$
Magnetic flux density	tesla	T	$kg\ s^{-2}\ A^{-1} = V\ s\ m^{-2}$
Luminous flux	lumen	lm	$cd\ sr$
Illumination	lux	lx	$cd\ sr\ m^{-2}$
Frequency	hertz	Hz	cycle per second
Customary temperature, t	degree Celsius	°C	$°C = K - 273 \cdot 15$

FRACTIONS AND MULTIPLES

Fraction	Prefix	Symbol	Multiple	Prefix	Symbol
10^{-1}	deci	d	10	deka	da
10^{-2}	centi	c	10^2	hecto	h
10^{-3}	milli	m	10^3	kilo	k
10^{-6}	micro	μ	10^6	mega	M
10^{-9}	nano	n	10^9	giga	G
10^{-12}	pico	p	10^{12}	tera	T
10^{-15}	femto	f			
10^{-18}	atto	a			

B

Conversion Factors for Determining Corrosion Penetration

A penetration rate of 1 g/m².day equals:

Metal	Density	mil/year (0·001 in/year) 1·44/density	µm/year 368/density
Aluminium	2·72	5·3	133
Beryllium	1·85	7·9	200
Brass	8·48	1·7	43
Cadmium	8·70	1·6	40
Cobalt	8·90	1·6	40
Copper	8·92	1·6	40
Germanium	5·35	2·7	68
Gold	19·3	0·74	19
Iron or steel	7·86	1·8	46
Lead	11·35	1·25	31
Magnesium	1·75	8·25	210
Molybdenum	10·2	1·41	36
Nickel	8·90	1.6	40
Niobium	8·55	1·68	43
Osmium	22·5	0·64	16
Palladium	11·4	1·25	32
Platinum	21·45	0·67	17
Silver	10·5	1·37	35
Tantalum	16·6	0·87	22
Tin	7·30	2·00	51
Titanium	4·55	3·2	81
Tungsten	19·3	0·74	19
Uranium	18·7	0·78	20
Zinc	7·15	2·00	51
Zirconium	6·45	2·25	57

Table of Metals Not Recommended for Contact With Various Chemicals

The metals considered are the following:

Aluminium: Al	Aluminium bronze:	Tin: Sn
Brass: Cu/Zn	Al/Cu	Copper: Cu
Mild Steel: m.s.	Bronze: Cu/Sn	Cast iron: Fe
Cupro-nickel: Cu/Ni	Duriron: Fe/Si	Nickel: Ni
Silver: Ag	Stainless steel: s.s.	Platinum: Pt
Zirconium: Zr	Tantalum: Ta	Titanium: Ti

Inclusion in brackets means that the metal in question is satisfactory at room temperature but not at elevated temperatures. Metals omitted are normally stable up to 100°C.

Acetic acid (dil.): (Al), Cu/Zn, Fe, Pb, m.s., Sn
Acetic acid (conc.): Cu/Zn, Fe, Pb, m.s., (s.s.), Sn
Acetic anhydride: Cu/Zn, (Cu/Sn), m.s. (s.s.), Sn
Acetone: (Pb), (m.s.)
Acetylene: Al/Cu, Cu, Cu/Sn, (Pb), Cu/Ni, (Sn)
Alkyl chlorides: Al, Cu/Zn, (Pb), (m.s.), (Sn)
Aluminium chloride: Cu/Zn, Fe, (Fe/Si), m.s., Ni, (Cu/Ni), (s.s.), (Sn)
Aluminium sulphate: Cu/Zn, Fe, m.s., (Ni), (Cu/Ni), (s.s.), (Sn)
Ammonia: Cu/Al, Cu/Zn, (Fe), Cu, Cu/Sn, (Pb), Ni, Cu/Ni, (Sn)
Ammonium chloride: Cu/Al, Cu/Zn, (Fe), Cu, Cu/Sn, (Pb), m.s., (Cu/Ni), (s.s.), (Sn)
Aniline: Cu/Al, Cu/Zn, Cu, Cu/Sn, (Pb), (Cu/Ni), (Sn)
Aqua regia: all with the exception of Ag and Ta; (Ti)
Benzene: (Pb), (m.s.)
Benzoic acid: Fe, Pb, m.s., (Sn)
Boric acid: (Al), Fe, m.s., (Cu/Ni),
Bromine: Cu/Al, Cu/Zn, Fe, Cu, Cu/Sn, (Fe/Si), (Pb), m.s., Ni, s.s., Sn, Ti
Calcium chloride: Cu/Zn, (Fe), (Pb), m.s., s.s., (Sn)
Carbon disulphide: m.s., (Ni), (Cu/Ni), (Sn)
Carbonic acid: Cu/Al, Cu/Zn, Fe, Cu/Sn, Pb, m.s., (Ni)
Carbon tetrachloride: Al, (Pb), (m.s.)
Chlorine (dry): Fe, Sn, Ti, (Zr)
Chlorine (wet): Al, Cu/Al, Cu/Zn, Fe, Cu, Cu/Sn, (Pb), m.s., Ni, (Cu/Ni), s.s., Sn, Zr
Chloroform: Al, (Fe), (Pb), (m.s.),
Chromic acid: Al, Cu/Al, Cu/Zn, Fe, Cu, Cu/Sn, (Fe/Si), m.s., Ni, Cu/Ni Ag, s.s.
Citric acid: Cu/Zn, Fe, m.s., (Cu/Ni),
Copper chloride: Al, Cu/Zn, Fe, Cu, Cu/Sn, (Pb), m.s., Ni, Cu/Ni, Sn, Zr

Copper sulphate: Al, Cu/Zn, Fe, Cu, Cu/Sn, (Pb), m.s., (Ni), (Cu/Ni), Sn, Zr
Ethanol: (Pb)
Ethyl acetate: (Pb), m.s. (Sn)
Fatty acids (higher): Cu/Zn, Fe, m.s., (Sn)
Ferric chloride: All except Ta, Ti; (Fe/Si), (Pt)
Ferrous sulphate: Al, Cu/Al, Cu/Zn, Fe, Cu, Cu/Sn, (Pb), m.s., Ni, (Cu/Ni), Ag, Sn
Fluorine gas (dry): Cu/Zn, Fe, (Pb), Fe/Si, Ta, Sn, Ti, Zr
Fluorine gas (wet): Al, Cu/Al, Cu/Zn, Fe, Pb, Fe/Si, m.s., s.s., Ta, Sn, Ti, Zr
Fluosilicic acid: all except Ni, Cu/Ni, Pt, Ag
Formalin solution (50%): (Al), (Cu/Zn), (Fe), (Pb), (m.s.), (Sn)
Formic acid: (Al), Fe, m.s., (Cu/Ni), (s.s.), Sn
Glycerol: (Pb), (m.s.)
Hexamine: Al, Cu/Al, Cu/Zn, Fe, Cu, Cu/Sn, (Pb), m.s. (Cu/Ni) (Sn)
Hydrobromic acid (conc.): all except Pt, Ta, Ti, Zr
Hydrobromic acid (dil.): ditto plus (Ni), (Cu/Ni), (Fe/Si)
Hydrochloric acid (conc.): all except Pt, Ta, Ti, Zr, (Cu/Al)
Hydrochloric acid (dil.): ditto plus Ag, (Fe/Si), (Ni), (Cu/Ni),
Hydrocyanic acid (conc.): Cu/Al, Cu/Zn, Cu, Cu/Sn, (Pb), (M.s.), (Cu/Ni), (s.s.), (Sn)
Hydrofluoric acid (60%): all except Cu/Ni, Pt, Ag; (Ni)
Hydrogen peroxide (conc.): Cu/Al, Cu/Zn, Fe, Cu, Cu/Sn, Fe/Si, Pb, m.s., Ni, Cu/Ni, Ag, (Sn), Ti
Hydrogen sulphide (wet): (Fe), (Pb), (Cu/Ni), Ag,
Lead acetate: Cu/Al, Cu/Zn, Fe, Cu, Cu/Sn, Fe/Si, Pb, m.s., (Cu/Ni), (Sn)
Mercuric chloride: all except Pt, Ta, Ag, Ti, Zr
Mercury metal: Al, Cu/Al, Cu/Zn, Cu, Cu/Sn, Pb, (Cu/Ni)
Naphthalene: (Pb), (m.s.)
Nickel chloride, nitrate and sulphate: Al, Cu/Al, Cu/Zn, Fe, Cu, (Pb), m.s., Sn
Nitric acid (dil.): all except Pt, s.s., Ta, Ti, Zr; (Fe/Si)
Nitric acid (conc.): all except Fe/Si, Pt, Ta, Ti, Zr; (s.s.)
Nitric acid (fuming): all except Fe/Si; (Al), (m.s.), (s.s.), (Ta), (Zr)
Oils (all kinds): (Pb), (m.s.)
Oxalic acid: (Al), (Cu/Zn), Fe, Pb, m.s., Ni, (Cu/Ni), (s.s.), (Ti)
Perchloric acid: all except Fe/Si, Pt, Ag, Ta, Ti, Zr
Phenol: (Cu/Ni), (Sn)
Phosphoric acid (conc.): Al, Cu/Al, Cu/Zn, Fe, Cu, Cu/Sn, (Pb), m.s., Ni, (Cu/Ni), (s.s.) Sn, (Ti), (Zr)
Phosphoric acid (dil.): (Al), (Cu/Al), Cu/Zn, Fe, Cu, Cu/Sn, m.s., Ni, (Cu/Ni), Sn
Phosphorus trichloride and pentachloride: Al, Cu/Al, Cu/Zn, (Fe), Cu, Cu/Sn, (Cu/Ni), s.s., Sn, (Zr)
Pyridine: Cu/Al, Cu/Zn, Cu, Cu/Sn, (Pb), (Cu/Ni),

Quicklime (CaO): (Al), (Pb), (Sn)

Sea-water: (Fe), (Pb), m.s., (Ni), (s.s.),

Silver nitrate: Al, Cu/Al, Cu/Zn, Fe, Cu, Cu/Sn, Pb, m.s., Ni, Cu/Ni, Sn

Sodium carbonate: (Al), (Pb), (Sn)

Sodium (and potassium) hydroxide: Al, Cu/Al, (Fe), (Fe/Si), Pb, (m.s.), Sn

Sodium chlorate: Cu/Zn, Fe, Pb, m.s., (Cu/Ni), (Sn)

Sodium chloride: Cu/Zn, m.s., (s.s.)

Sodium hypochlorite: all except Fe/Si, (Pb), Pt, (Ag), Ta, Ti, Zr

Sodium peroxide: Al, Cu/Al, Cu/Zn, Cu, Cu/Sn, Pb, m.s., (Cu/Ni), (s.s.), Sn

Sodium sulphate: (Pb), m.s.

Sodium sulphide: Al, Cu/Al, Cu/Zn, Cu, Cu/Sn, (Fe/Si), m.s., Ag, (s.s.), Sn, (Ti)

Sulphur: Cu/Al, Cu/Zn, (Fe), Cu, Cu/Sn, Pb, Ag, (Sn)

Sulphur dioxide (dry): (Fe), (Cu/Ni)

Sulphur dioxide (wet): Cu/Zn, Fe, Cu, Fe/Si, Cu/Sn, Pb, m.s. Ni, Cu/Ni, (s.s.), Sn

Sulphur trioxide: Al, Fe, Cu/Ni, s.s., Sn, Ti, Zr

Sulphuric acid (dil.): Al, Cu/Zn, Fe, Cu/Sn, m.s., Ni, (Cu/Ni), s.s., Sn

Sulphuric acid (conc.): Al, Cu/Al, Cu/Zn, (Fe), Cu, Cu/Sn, (Pb), (m.s.), Ni, Ag, (s.s.), Sn, Ti, Zr

Sulphuric acid (fuming): (Al), Cu/Al, Cu/Zn, Cu, Cu/Sn, Pb, (m.s.), Ni, Cu/Ni, Ag (s.s.), Sn, Ti, Zr

Sulphur chlorides: all except Pt, Ta, (Fe), (Pb), (Cu/Ni)

Tannic acid: Fe, Pb, m.s., (Sn)

Tartaric acid: Fe, m.s.,

Tin salts: all except Pt, Ta, Ti, Zr (Fe/Si), (Ag)

Trichloroethylene: (Fe), (Pb), (m.s.)

Zinc salts: Al, Cu/Al, Cu/Zn, Fe, Cu, Cu/Sn, (Fe/Si), (Pb), m.s., s.s., Sn

Chapter 1   The Nature of Corrosion

One can define the term 'corrosion' as the conversion of a metal to a metallic compound. This means that the essential metallic qualities of strength, elasticity, ductility, etc., are lost, and instead substances are being produced which are extremely poor with regard to these properties. It is estimated that the financial loss in the United Kingdom alone caused by corrosion is approximately £600 million ($1,450 million) each year. Corrosion causes wastage of materials, expensive failure of capital equipment, as well as high maintenance costs. Often it is the failure of quite small components which make a large capital installation useless; a modern equivalent to the old saga of the battle that was lost due to the lack of a horseshoe nail!

Metals are reduced from their compounds during various metallurgical processes. These processes are, in nearly all cases, endothermic in that a large quantity of energy has to be expended in the form of heat from fuels to make the transformation:

$$\text{metallic ore} \longrightarrow \text{metal}$$

While the metallic ore has negligible strength properties, the metals mostly have good qualities in this respect, and can therefore be used for the manufacture of the essential items that our civilisation needs. Because the metals contain more free energy than the corresponding metallic oxides, hydroxides, carbonates, etc., from which they originate, there is an intrinsic tendency for the metals to revert to such compounds and give off energy in the process. Fortunately corrosion, or the conversion of a metal back into its oxide, etc., is a surface chemical reaction only, and there are, therefore, a number of ways of slowing down or even stopping this reaction.

It is, however, absolutely necessary to have a comprehensive understanding of the precise mechanism of the corrosion processes involved. In most cases protective films are produced, which are virtually impervious to atmospheric gases, which then rapidly inhibit further corrosion. In other cases it is possible to inhibit corrosion by stopping the cathodic reaction, which means that the energy produced in the form of electricity by the corrosion process is prevented from being discharged. Because of the application of the Le Chatelier principle, which states that the progress of a reaction is influenced by a constraint which is put upon it, the corrosion reaction is slowed down or stopped.

1.1 Surface corrosion of metals in dry air

When metals such as iron or copper are heated in air, they tend to form oxide films. It is true that the oxide film has a lower energy content than the pure metal and therefore one would expect such a film to be formed spontaneously even in the cold. In fact there is an energy barrier which has to be overcome before oxidation can take place. Heating the metal serves to overcome this energy barrier (Fig. 1.1).

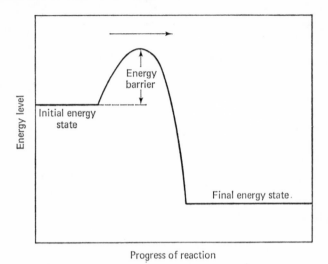

Fig. 1.1. Energy barrier in a chemical reaction

The reaction which takes place with respect to iron is the following:

$$3Fe + 2O_2 \rightarrow Fe_3O_4$$

It should, however, be pointed out that such a reaction does not give an accurate picture of what happens, because Fe_3O_4 is really a double compound of FeO and Fe_2O_3.

As the oxide is formed at the surface of the iron or other metal it constitutes a barrier to further reaction and therefore the reaction rate is slowed down.

The rate of oxidation of many metals follows the parabolic law of oxidation or (Fig. 1.2).

$$y = (At + B)^{\frac{1}{2}} \tag{1.1}$$

where y is the thickness of the film in millimetres, t the time in seconds, and A and B constants depending upon the nature of the metal, the atmosphere to which it is exposed, the external temperature and a number of

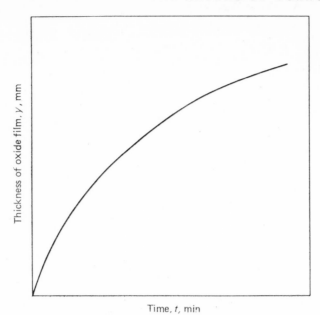

Fig. 1.2. Parabolic oxidation law: $y = (At + B)^{\frac{1}{2}}$

other factors. The parabolic law is obeyed by iron at temperatures above 200°C and by copper at temperatures above 800°C.

Iron, which is oxidised in air at a temperature below 200°C, copper at temperatures below 800°C as well as zinc and aluminium at all temperatures obey the logarithmic law of oxidation (see Fig. 1.3). In this

$$y = A \log_{10} (Bt + C) \qquad (1.2)$$

A, B and C are constants.

In all cases where a logarithmic law of oxidation applies, the rate of oxidation is fast at first, but slows down very quickly to an extremely low value. In the case of metals where a parabolic law is obeyed, a slow-down of oxidation also takes place, but it is not quite as pronounced.

There are some metals where the oxide film is so porous that it offers virtually no resistance to the passage of oxygen to the surface of the metal. In such cases the film increases in thickness at a constant rate, or in other words a rectilinear law of oxidation is obeyed (Fig. 1.4). In such a case the equation becomes:

$$y = At + B \qquad (1.3)$$

A and B again being constants.

The rectilinear law applies to the atmospheric oxidation of titanium, sodium, potassium, calcium and magnesium.

In certain cases oxidation proceeds for a time according to one of the laws, changes over to another and may then come back to the first. This is often the case when a thin protective oxide is formed, which starts to

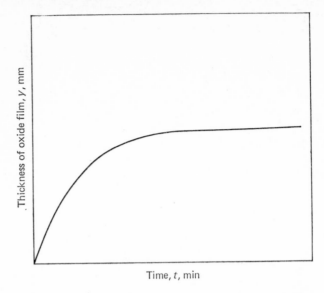

Fig. 1.3. Logarithmic oxidation law: $y = A \log_{10} (Bt + C)$

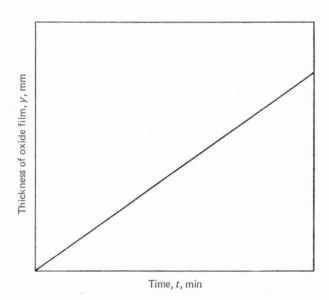

Fig. 1.4. Rectilinear oxidation law: $y = At + B$

break up after it has reached a given thickness. A typical example of this is copper, which starts to oxidise in accordance with a parabolic law, but owing to the repeated break-up of the surface films the various parabolic graphs become discontinuous so that the net effect appears to follow a rectilinear type of oxidation (Fig.1.5).

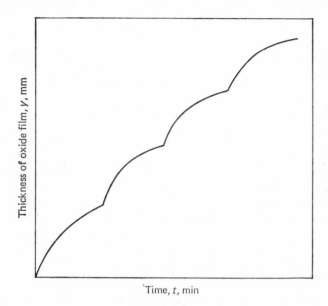

Fig. 1.5. Discontinuous parabolic law

1.2 The effect of mercury on aluminium

Aluminium normally oxidises in air according to a parabolic law and after rapid initial oxidation the rate of further oxidation becomes very slow. If, however, the surface of the aluminium becomes amalgamated, i.e. contaminated with mercury so that an aluminium-mercury couple is formed, the film of oxide produced has a negligible resistance to air and the oxidation of the aluminium now takes place according to a rectilinear law. Actually, the rate of oxidation becomes so rapid that a white flocculent film of aluminium oxide is formed at a visible rate. The aluminium becomes very warm because the reaction is strongly exothermic:

$$4Al + 3O_2 \rightarrow 2Al_2O_3$$

Amalgamation of the aluminium surface can be caused either by direct contact with mercury metal or by immersion in a solution of mercuric chloride or other mercuric salt.

1.3 Reasons why many oxide films inhibit further oxidation

Contrary to common belief, metallic ions together with the electrons which have been given off during the ionisation process migrate through oxide films to the outside. Oxygen gas at the surface of the oxide film is changed to oxygen ions by the absorption of electrons:

$$O_2 + 4e \rightarrow 2O^{2-}$$

the electrons having been conducted through the oxide film. These oxygen ions are formed at the surface of the oxide film and react with the metal ions which also diffuse through the film to form more metallic oxide. The most important factor and rate-determining step in corrosion is the speed with which electrons can pass through the oxide film to form the oxygen ions. A thicker oxide film has a greater electrical resistance than a thin one, and therefore the speed at which the electrons are supplied to the oxygen is reduced. Sometimes the nature of the oxide film also governs the speed of travel of the electrons. For example, if copper is coated with cupric oxide (CuO), as it is if plenty of oxygen is present, the rate of corrosion is more rapid than if it is coated with cuprous oxide (Cu_2O) in an oxygen-starved atmosphere. The reason is simply that Cu^{2+} ions conduct electricity better than Cu^+ ions.

1.4 Prevention of surface oxidation by alloying

The rate of surface oxidation of a metal, where the oxide film has a high electric conductivity and thus causes the formation of thick oxide films, can often be reduced by alloying. A typical example of this is the alloying of steel with chromium; adding between about 13 and 18% of chromium to steel means that a surface film very rich in Cr_2O_3 is formed. In contrast to iron oxides, Cr_2O_3 has a very low electrical conductivity; oxidation of the metal is thereby very strongly reduced. Steels are only "stainless" if the amount of chromium and/or other metal forming an oxide of low electrical conductivity is sufficient to inhibit the formation of iron oxide.

1.5 Corrosion in solution

The most common form of corrosion is not straight surface oxidation, as described above, but a rather complex electrolytic reaction involving the flow of electric current from the anodic portion of the corrosion cell, which oxidises, to the cathodic section, which disposes of the electrons produced.

1.6 Elementary electrochemistry

An electrolytic cell consists of two electrodes immersed in a solution containing positive and negative ions (Fig. 1.6).

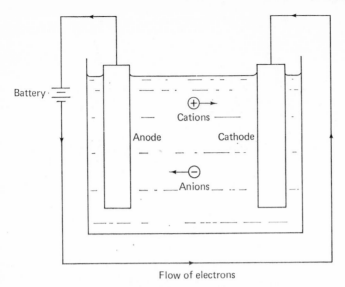

Flow of electrons

Fig. 1.6. Conventional representation of an electrolytic cell

If the two electrodes are connected to each other via a wire or other electrical connection, electrons can pass from one to the other.

The general convention of naming is as follows.

The *anode* is the electrode to which the *anions* or negatively charged ions travel. The *cathode* is the electrode to which the *cations* or positively charged ions travel. The electrons flow always from the anode to the cathode. In an electrolytic cell a battery is included in the circuit which serves to withdraw electrons from the anode and transfers electrons into the cathode.

A corrosion cell is simply an electrolytic cell where this process of electron transfer proceeds spontaneously, the driving force being the difference in electric potential between the anode and the cathode. The energy required to do the work of 'conveying' the electric current is obtained from the difference between the free energy in the metal and that in the corrosion product (of the metal). With all metals prone to corrosion the metal has a much higher free energy content than the products derived from it (Fig. 1.7).

With an electrolytic cell one tries to deposit a metal from a metal salt. This means a step from a material with a lower free energy to a material with a higher free energy. Consequently, energy in the form of the driving power of a battery or generator is required to make the electrons flow.

However, the convention of the terms 'anode' and 'cathode' are the same in corrosion cells as they are in electrolytic cells. Electrons always flow from the anode to the cathode, and never vice versa.

The properties of electrolytic cells are governed by Faraday's law which states that 1 gramme-equivalent of any substance is transferred from the anode or cathode by a quantity of electricity of 96,500 C, where 1 C = 1 A.sec.

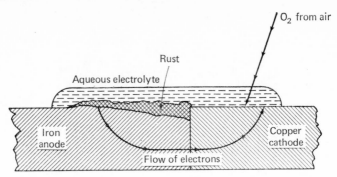

Fig. 1.7. A typical corrosion cell. There is a spontaneous flow of electrons from the
iron anode to the copper cathode

Faraday's law applies to corrosion cells. For each 96,500 C which pass
between the anode and the cathode of a corrosion cell, 1 gramme-equivalent
of the metal is converted into corrosion products. It is therefore necessary
to ensure that the current flow is kept down to a minimum or stopped
altogether. The lower the current passing between anode and cathode, the
slower the rate of corrosion will be.

1.7 Reversible cells

A typical reversible cell can be represented by the Daniell cell (Fig.1.8):

$$\text{Zinc metal}/MZn^{2+}/MCu^{2+}/\text{copper metal}$$

MZn^{2+} indicates a solution of 1 g.mole of a zinc salt, such as $ZnSO_4$, per
litre, and MCu^{2+} indicates 1 g.mole of a copper salt, such as $CuSo_4$, per litre.

The conventional way of depicting an electrolytic cell is to position the
anode always on the left-hand side, and the cathode on the right. On the
left-hand side of this cell, zinc therefore goes into solution as follows:

$$Zn \rightarrow Zn^{2+} + 2e$$

or, in other words, using the definition that oxidation increases the electropositive valency, the zinc metal is being oxidised. On the left-hand side of
the cell copper comes out of solution:

$$Cu^{2+} + 2e \rightarrow Cu$$

or, in other words, the cupric ion is being reduced to the metal. In such a
cell there is a positive e.m.f. of 1·10 V which makes the electrons flow and
causes the reaction. The reaction can be stopped by the application of this
voltage, and reversed if a voltage in excess of this is applied in a direction
opposite to the natural difference in e.m.f. This e.m.f. of 1·10 V which is
exactly equal to the natural difference in potential between the anode and
the cathode of a cell, is called the reversible cell potential. It depends on
the following factors:

1 The e.m.f. of the electrode materials under standard conditions which involve the presence of 1 g.mole of the ions of substance per litre, as well as a pressure of the electrode material of 1·013 bar if this should be gaseous.
2 The concentration of oxidised anodic material round the anode in gramme moles per litre.
3 The concentration of material to be reduced round the cathode in gramme moles per litre.

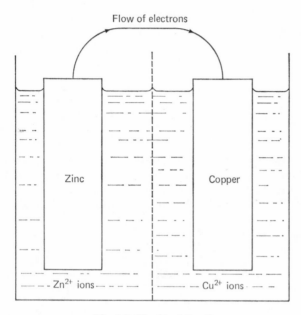

Fig. 1.8. The Daniell cell

1.8 Reversible cell potential and thermodynamics

In order to have in an electrolytic cell a cell reaction that is spontaneous, i.e. one that proceeds without an external agency such as the battery in an electrolytic cell, there must be a negative difference of free energy between the electrodes.

Let $-\Delta F$ (J/mole) be the difference in Gibbs free energy of such a cell. $\mathbf{F}$ (C/mole) = the Faraday constant, ΔE (V) the difference of e.m.f. between anode and cathode and n = the valency of the ions concerned. Then

$$-\Delta F = n\Delta E\mathbf{F} \qquad (1.4)$$

Now consider a typical reversible reaction:

$$yA + ZB... \rightleftharpoons pX + qY... \qquad (1.5)$$

According to basic chemical thermodynamics:

$$-\Delta F = -\Delta F^\circ - RT \log_e \frac{A^y \times B^z \ldots}{X^p \times Y^q \ldots} \qquad (1.6)$$

or

$$\Delta E \, n\mathbf{F} = \Delta E^\circ \, n\mathbf{F} - RT \log_e \frac{A^y \times B^z \ldots}{X^p \times Y^q \ldots} \qquad (1.7)$$

where A, B, X and Y (g.mole/litre) represent the activities of the various components, R the universal gas constant (or 8·314 J/K mole), T (K) the temperature and ΔE° the free energy difference when standard conditions prevail, i.e. when the various components are present in concentrations of 1 g. mole/litre throughout.

Therefore dividing Eq. (1.7) by $n\mathbf{F}$ throughout we get:

$$\Delta E = \Delta E^\circ - \frac{RT}{n\mathbf{F}} \log_e \frac{A^y \times B^z \ldots}{X^p \times Y^q \ldots} \qquad (1.8)$$

where ΔE is the reversible cell potential difference between the anode and cathode of the corrosion cell concerned and ΔE° the difference between the standard electrode potentials of the electrode materials given in the table (see p 14). It makes things easier to standardise all reactions to one temperature, namely 25°C or 298·15 K and to include the conversion to normal or base 10 logarithms.

Let

$$\frac{A^y \times B^z \ldots}{X^p \times Y^q \ldots} = M$$

Then $(RTn\mathbf{F}) \log_e M$ becomes $(0·0592/n) \log_{10} M$ at 25°C.

The cell reaction taking place in the electrolytic cell is also rather simpler than in the general case mentioned before. At the anode the following reaction takes place:

$$A_{(solid)} \rightleftharpoons A^{n+} + ne$$

while at the cathode the following reaction takes place:

$$B^{m+} + me \rightleftharpoons B_{(solid)}$$

To account for the fact that the valencies involved in the anode and cathode reactions may differ, we separate the E values:

$$\Delta E = [E^\circ(A) - (0·0592/n) \log_{10} A^{n+}] \\ - [E^\circ(B) - (0·0592/m) \log_{10} B^{m+}] \qquad (1.9)$$

where n and m are the valencies of the anode and cathode ions, respectively.

1.9 Activities

The values dealing with A^{n+} and B^{m+} are given as activities where the activity is defined as the molar concentration c multiplied by the activity coefficient γ. (*Molar* means 1 g.mole/litre of solution and *molal* means 1 g.mole/kg of solvent.) The activity coefficient γ varies considerably with the nature of the solution involved and also the nature of the ions concerned. The value of γ approaches unity with very dilute solutions, and achieves a minimum as the solution is concentrated, rising once more after this critical concentration is exceeded. Table 1.1 shows the way in which γ varies with respect to different electrolytes and different concentrations.

TABLE 1.1 Mean molar ionic activity coefficient γ of pairs of ions at 25°C*

Molar concentration, g.mole/litre	HCl	NaCl	$CaCl_2$	$ZnCl_2$	$ZnSO_4$
0·001	0·966	0·966	0·888	0·881	0·734
0·005	0·930	0·928	0·789	0·767	0·477
0·01	0·906	0·903	0·732	0·708	0·387
0·05	0·833	0·821	0·584	0.502	0·202
0·1	0·798	0·778	0·524	0·502	0·148
0·5	0·769	0·679	0·510	0·376	0·063
1·0	0·811	0·656	0·725	0·325	0·044
2·0	1·011	0·670	1·554		0·035
3·0	1·31	0·719	3·380		0·041

* S. Glasstone: *Textbook of physical chemistry.*

As can be seen from the table the mean activity coefficient of an ion at a given concentration depends not only upon the nature of the ion itself, but also upon the nature of other ions present. In corrosion cells, where the number of different dissolved ions in the electrolyte can be large, e.g. where sea-water is used as an electrolyte, it may be difficult to evaluate activities from first principles. Empirical methods are used instead.

1.10 Similarity between the Leclanché cell and a corrosion cell

One can visualise the way a corrosion cell works by comparing it with a Leclanché cell, or torch battery (see Fig. 1.9). This consists of a central carbon rod surrounded by a bag filled with manganese dioxide, which serves as an oxidising agent. The electrolyte of the cell is an aqueous solution of ammonium chloride and the external shell consists of zinc. The process by which the reaction takes place is as follows. The zinc on the outside and the carbon rod in the centre have a differing e.m.f. and

c

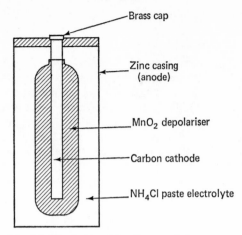

Fig. 1.9. The Leclanché cell or torch battery

in consequence the zinc becomes the anode of the electric cell and the carbon rod the cathode. At the anode the following reaction takes place:

$$Zn \rightarrow Zn^{2+} + 2e$$

The Zn^{2+} then reacts with the Cl of the electrolyte:

$$Zn^{2+} + 2Cl^- \rightarrow ZnCl_2$$

while the NH_4^+ ions reacts with water:

$$NH_4^+ + H_2O \rightarrow NH_4OH + H^+$$

or, more correctly,

$$NH_4^+ + 2H_2O \rightarrow NH_4OH + H_3O^+ \text{ (solvated hydrogen ion)}$$

The hydrogen ion moves towards the carbon cathode.

However, it is impossible for hydrogen to be discharged at the carbon surface due to various effects which are called polarisation or overvoltage (which will be described later). This means that if there were no manganese dioxide (depolariser) to oxidise the hydrogen ions into water according to the following reaction:

$$2MnO_2 + 2H^+ + 2e \rightarrow Mn_2O_3 + H_2O$$

the cell reaction would stop. The polarisation caused by the accumulated hydrogen ions would produce a back-e.m.f. sufficient to counteract the forward-e.m.f. due to the difference in electrical potential between the zinc and carbon.

If the anode and the cathode of the cell are connected by a wire, the electrons produced on the zinc anode pass through it, to be discharged at the cathode, thus enabling the hydrogen ions to form water with some of the oxygen contained in the MnO_2. If the connection between the anode and cathode is interrupted, the entire cell action stops as the hydrogen ions accumulate on the cathode side to produce a back-e.m.f. which rises until it equals the forward-e.m.f. of the cell.

The mechanism of electrolytic corrosion is very similar to the process just described. For corrosion to take place the following conditions are required:

1 There must be a difference in electrical potential between the anode and the cathode. This means that for metals the anode and cathode must:
a Consist of different metals.
b Consist of different alloys of the same metal.
c Have different concentrations of oxygen or electrolyte around them.

2 There must be an electrolyte present; this need only be pure water, as even this is slightly ionised. Corrosion reactions are accelerated considerably when the electrolyte consists of a salt solution or similar, so that there is a much higher ionic concentration.

3 If the pH of the electrolyte is less than 3, i.e. if the solution is very acid, it is often possible for hydrogen to be evolved directly. If, however, the pH of the electrolyte is in excess of this value, corrosion cannot take place unless there is some means present of oxidising the hydrogen ions on the cathode side. Normally this oxidising agent is simply atmospheric oxygen. However, in such cases as the corrosion that takes place at the bottom of oil tanks, oxidation of the hydrogen ions is carried out by the dissolved oxygen. If air or oxygen is excluded from the cathode side of a corrosion cell, the reaction stops.

4 An electric connection must exist between the anode side of the corrosion cell and the cathode side to enable the electrons produced at the former to be transferred to the latter where they are needed to dispose of the hydrogen ions produced during the reaction.

1.11 Potential differences in corrosion cells

To determine which of the two metals in a pair is likely to become the anode and corrode, and which is likely to remain the cathode, reference is made to the standard electrochemical series. This gives the e.m.f. in volts at 25°C in relation to hydrogen. Table 1.2 deals with a number of common metals.

In any galvanic cell, the e.m.f. between the two electrodes at 25°C is given by

$$\Delta E = [E° (A) - (0{\cdot}0592/n) \log_{10} A^{n+}]$$
$$- [E° (B) - (0{\cdot}0592/m) \log_{10} B^{m+}] \qquad (1.10)$$

where ΔE is the potential difference between anode and cathode at reversibility, i.e. when no current is flowing, n and m are the valencies of the

A^{n+} and B^{m+} ions, respectively, and A^{n+} and B^{m+} are represented in the equation by their activities.

If ΔE is a positive value, then metal A becomes the anode of the corrosion cell and corrodes, and metal B becomes the cathode and is unattacked. The latter serves to dispose of the electrons produced in the oxidation reaction which takes place at the anode.

If ΔE has a high value, we can expect a fast start to corrosion processes, although the subsequent rate of corrosion is markedly affected by the various polarisation processes that then take place. These drastically lower ΔE when a current is actually passing.

TABLE 1.2

Electrochemical series of common metals

$Na \rightarrow Na^+ + e$	$E° = + 2·71$ V
$Mg \rightarrow Mg^{2+} + 2e$	$E° = + 2·37$ V
$Al \rightarrow Al^{3+} + 3e$	$E° = + 1·66$ V
$Ti \rightarrow Ti^{2+} + 2e$	$E° = + 1·63$ V
$Zn \rightarrow Zn^{2+} + 2e$	$E° = + 0·763$ V
$Cr \rightarrow Cr^{3+} + 3e$	$E° = + 0·74$ V
$Fe \rightarrow Fe^{2+} + 2e$	$E° = + 0·440$ V
$Ni \rightarrow Ni^{2+} + 2e$	$E° = + 0·250$ V
$Sn \rightarrow Sn^{2+} + 2e$	$E° = + 0·136$ V
$Pb \rightarrow Pb^{2+} + 2e$	$E° = + 0·126$ V
$H_2 \rightarrow 2H^+ + 2e$	$E° = \quad 0$
$Cu \rightarrow Cu^{2+} + 2e$	$E° = - 0·337$ V
$Ag \rightarrow Ag^+ + e$	$E° = - 0·80$ V
$Pt \rightarrow Pt^{2+} + 2e$	$E° = - 1·20$ V
$Au \rightarrow Au^{3+} + 3e$	$E° = - 1·50$ V

Metals that are particularly harmful to iron or steel when in contact with it are those which not only possess a very low $E°$ value, but are also virtually insoluble in the usual corrosive agents surrounding the iron/metal couple. Consequently, one of the most harmful is copper.

It is also possible for metals or alloys that have intrinsically higher $E°$ values than iron to be cathodic when in contact with iron. Aluminium and chromium are cases in point. They become covered by passive films so that the metal is never exposed to the electrolyte at all, and in consequence both these metals behave like oxygen electrodes. Instead of becoming anodes, as their position in the electrochemical series would suggest, they often behave like cathodes, accelerating the corrosion of any iron or steel in contact with these metals.

On the other hand, when tinned steel is in contact with food on the inside of tin cans, the formation of tin complexes with food acids actually makes the tin anodic with respect to steel, thereby protecting the latter. This does not happen when tin is exposed to the atmosphere. In such circumstances the tin is cathodic, causing rapid corrosion of the steel, if the steel should at any time be exposed to the atmosphere owing to a scratch, etc.

1.12 Practical electrochemical series

Owing to such effects as passivity and polarisation (which will be described later), the electrochemical nature of metals varies with the type of electrolyte to which they are exposed. The list below gives the relative position of common commercial alloys when in contact with sea-water. This also gives a rough indication regarding their relative position in normal industrial atmospheres. The list is written with the most anodic at the top, and one can therefore assume that in most cases, where couples are made up of two members of the list, the one closer to the top of it will corrode while the other will act as the cathode. This is, however, by no means true in all cases and this list should be taken as an approximate guide only.

Magnesium, magnesium alloys, zinc, pure aluminium, cadmium, duralumin, mild steel, wrought iron, cast iron, 50/50 lead–tin solder, 18/8 stainless steel (active), lead, tin, bronzes, nickel (active), Iconel (Ni/Cr/Fe alloy), (active), brasses, copper, silicon bronze, nickel (passive), Iconel (passive), 18/8 stainless steel (passive), Cr/Ni/Mo stainless steel (passive), titanium.

As can be seen, the existence of a passive (usually a coherent film of oxides, etc.) coating on the surface of the metals makes an enormous difference with regard to their position in the electrochemical series. This means, of course, that while fresh· aluminium, which is fairly active, has a distinctly anodic position to iron and steel, the same is not true of aluminium where a good passive film has formed. In the same way, stainless steel, which is still active, will corrode when in contact with copper, bronzes and brasses. When stainless steel has been exposed to the atmosphere, it becomes one of the most cathodic metals in common use.

1.13 The effect of polarisation

The driving force for electrochemical corrosion is the difference in electrical potential between the anode and the cathode of the corrosion cell. But during the actual process of corrosion when a current is flowing, the e.m.f. is often quite different from the so-called 'reversible' value which applies when no external current is being drawn from or supplied to the cell. Both the anode and the cathode of a corrosion cell can exhibit the phenomenon of polarisation, or 'overvoltage' as it is often called (see Fig. 1.10).

ACTIVATION POLARISATION

Let us consider a case where a metal dissolves to form an ion:

$$M \rightarrow M^{n+} + ne$$

This is the anodic reaction and signifies that the metal goes into solution as a metallic ion, i.e. that the metal corrodes. On the other hand, the reaction that takes place at the cathode is

$$M^{n+} + ne \rightarrow M$$

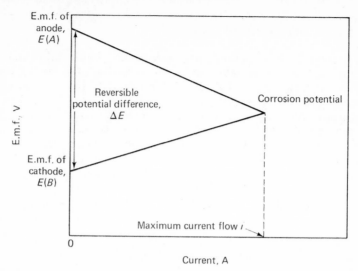

Fig. 1.10. The effect of anodic and cathodic polarisation on the corrosion current i

where M in this case is either another metal, or as is usual with corrosion cells, it represents the gas hydrogen.

Neither of these two reactions is reversible. This means that before the reaction can occur, a given energy barrier must be overcome. As an analogy, we may consider a lump of hard coke. Coke, when it burns, gives off a lot of heat. Before this heat can be released it is necessary to raise the temperature of the lump of coke above its ignition temperature.

For example, in an electrolytic cell in which the anode consists of iron and the cathode of nickel, no corrosion current passes since the polarisation of the nickel swamps the 0·19 V difference in basic potential $E°$ between the two metals. Equally, no corrosion can take place when hydrogen is evolved at the cathode **in the absence of air**, unless the hydrogen ion concentration of the solution is very high, i.e. the hydrogen ion concentration must be in excess of 10^{-3} g.ions/litre, or in other words the pH must be below 3. Corrosion when the pH is above 3 is then similar to the likelihood of setting fire to a large lump of high density coke with a match. In neither case is the energy sufficient to overcome the barrier.

Activation polarisation varies enormously with different materials. It is high with metals such as nickel and iron, and also for gases, but low for silver and copper.

CONCENTRATION POLARISATION

When metal ions dissolve at the anode, the vicinity of the anode gets swamped with anode ions and, therefore, there is a marked change in the so-called concentration potential, which is the

$$(0·0592/n_A) \log_{10} A^{n+}$$

part of the expression

$$E(A) = E°(A) - (0.0592/n_A) \log_{10} A^{n+} \qquad (1.11)$$

where $E°$ is the effective e.m.f. of the anode on the arbitrary scale which gives hydrogen at a pressure of 1·013 bar in the presence of 1 g.mole of hydrogen ions per litre, a voltage figure of zero. The effect is, of course, that $E(A)$ is lowered drastically.

Similarly, as metal ions, hydrogen ions, etc., are being discharged at the cathode, the vicinity is being denuded of these ions so that

$$(0.0592/n_B) \log_{10} B^{m+}$$

falls, with the result that

$$E(B) = E°(B) - (0.0592/n_B) \log_{10} B^{m+} \qquad (1.12)$$

increases.

As $\Delta E = E(A) - E(B)$, the net result is that the potential difference between the electrodes is drastically reduced. Concentration polarisation, unlike activation polarisation, increases with time.

As the current flowing between anode and cathode increases, $E(A)$ falls and $E(B)$ rises, until the values of the two become identical. This means that at a given current the potentials of the anode and cathode become zero, and consequently there is a limit to the magnitude of the current that can pass with a given electrolytic cell. As the rate of corrosion depends on the magnitude of this current, high polarisation obviously slows down the rate of corrosion.

The maximum rate of corrosion w is given by the following equation:

$$w = Wi/\mathbf{F} \text{ grammes metal dissolved/second} \qquad (1.13)$$

where W is the equivalent weight of the metal concerned, i (A) the maximum current passing and $\mathbf{F}$ equals Faraday's constant or 96,500 C. The maximum current passing can be read off from the graph (Fig. 1.10) by considering the value corresponding to the e.m.f. at which the values $E(A)$ and $E(B)$ come together. This value is given the name: *corrosion potential.*

Systems are designed to limit corrosion in such a way that the corrosion potential is reached at as low a current flow as possible. Making the anode passive (a technique that will be described later) steps up the polarisation of the anode considerably, and very much reduces i, which is the maximum current that passes, and hence the rate of corrosion (Fig. 1.11). Anodic inhibition, which results in the formation of a product that is only sparingly soluble in water, close to the point where solution takes place, has a similar effect of increasing anodic polarisation.

Cathodic polarisation is mainly due to the considerable activation polarisation of hydrogen; this varies with the different metals concerned. As can be seen from Fig. 1.11, the cathodic polarisation graph for the reaction

$$2H^+ + 2e \rightarrow H_2$$

is lower for iron than for zinc.

Although ΔE is much higher for a zinc–hydrogen pair than for an iron–

hydrogen pair, iron corrodes more rapidly than zinc in a dilute acid solution. The reason is obvious: the current corresponding to the corrosion potential of a zinc–hydrogen cell is less than that which corresponds to the corrosion potential of an iron–hydrogen cell. Because of the high degree of cathodic polarisation of the process,

$$2H^+ + 2e \rightarrow H_2$$

the corrosion current passing is quite insignificant when the hydrogen ion concentration at the cathode side of the corrosion cell is well below 10^{-3} g.ions/litre (the pH is above 3). In the absence of oxygen or another oxidising agent at the cathode side, virtually no corrosion can take place. This explains why steel objects that have lain at the bottom of the sea for decades or even centuries, show very little signs of corrosion.

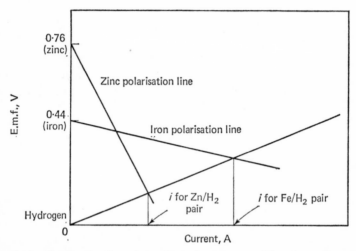

Fig. 1.11. Graph showing that although the reversible electrode potential of ZN Zn/H_2 is greater than that of Fe/H_2, the latter corrodes more rapidly because the polarisation graph is shallower

1.14 Cathodic oxidation of hydrogen ions

In neutral and near neutral solutions the cathode process involves the absorption of oxygen from the air. In such circumstances an oxygen–hydroxyl or O^2/OH^- cell that has a very low $E(B)$ value is set up. At 25°C this works out at

$$E(B) = -0.401 - 0.0148 \log_{10} \{a\,(O_2)/[a\,(OH^-)]^4\} \qquad (1.14)$$

where $a(O_2)$ equals the activity of oxygen or approximately the pressure of oxygen in atmospheres upon the cathode (1 atm equals 1·013 bar) and $a(OH^-)$ equals the activity of the hydroxyl ion or approximately the ionic concentration of OH^- in gramme ions per litre.

With a neutral salt solution and an oxygen activity of 1·013 bar, the $E(B)$ value equals -0.814 V, i.e. a rather lower negative value than silver which has an $E°$ value of $-0·80$ V.

It becomes obvious that an oxygen–hydroxyl cathode is an extremely potent one and readily induces corrosion of almost any metal at the anode of such a cell. This is the reason why a metal such as aluminium, which has an $E°$ value higher than iron, does not protect iron when in contact with it. The reverse, in fact, takes place. The aluminium surface gets covered by a completely impervious film of aluminium oxide, which acts as an oxygen electrode. Consequently, the reversible ΔE figure between the aluminium and the iron anode becomes close to $0·44 - (-0·814)$ or $1·254$ V (which is even worse than the E value that exists in an iron-copper couple) which is equal to $0·44 - (-0·337) = 0·777$ V. On the other hand, the

$$O_2 + H_2O + 4e \rightarrow 4OH^-$$

reaction is one which is considerably affected by concentration polarisation, and consequently the maximum corrosion current that can pass is limited. The rate at which the reaction can take place is governed by the rate at which oxygen can be supplied to the cathode surface and the rate at which the hydroxyl ions produced can be removed from the vicinity of the cathode. Stirring or other agitation of the cathode side of the cell increases the rate of corrosion manifold.

From all this it becomes obvious that a high degree of polarisation cannot prevent corrosion absolutely, as at zero current there is no polarisation, and therefore the natural ΔE values apply. These depend only on the nature and quantity of electrode material and the ions contained in the solution. As soon as a current passes, the ΔE value falls and in consequence the current itself, which depends on

$$i = \Delta E / R \text{ (Ohm's Law)} \tag{1.15}$$

is affected. As the rate of corrosion is directly proportional to the current i passing between anode and cathode, corrosion can be reduced by two methods:

1 The ΔE value can be reduced by seeking conditions of high polarisation.
2 The resistance of the path between the anode and the cathode of the cell should be as high as possible. This can be done either by insulating the two electrodes physically from each other, or by coating them with paint, plastics films etc.

1.15 Passive films

A method of increasing the polarisation properties of electrodes is to make them passive (see Fig. 1.12). For example, if iron is treated with concentrated nitric acid, the surface becomes covered with an extremely thin

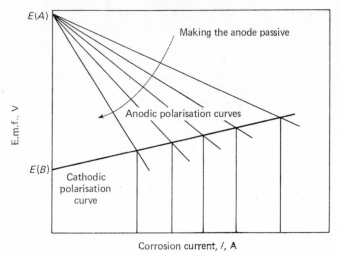

Fig. 1.12. Graph showing marked reduction in the corrosion current on passivation of the anode

layer of ferric oxide, and in consequence it becomes increasingly difficult for the following reaction to take place:

$$Fe \rightarrow Fe^{2+} + 2e$$

The anodic polarisation increases to a level permitting only minute corrosion currents to pass. If the extremely thin ferric oxide layer should be broken the passivity of the iron is partly destroyed at that point.

Stainless steels can be made passive by exposing them to air and provided they are kept in atmospheres which are even slightly oxidising, any scratched passive film on the surface is readily healed.

1.16 Galvanic corrosion

Galvanic corrosion takes place whenever two metals with different $E°$ values are in contact with each other. It is the form of corrosion which takes place when iron is in contact with a metal which has an $E°$ value of less than $+0.44$ V. In such circumstances the iron becomes the anode and goes into solution as Fe^{2+} ions. Hydrogen ions from the electrolyte pass to the cathode, and the electrons released during the anodic reaction are conducted to the cathode via the metallic contact of the two metals. At the cathode side, the reactions that take place are

$$O_2 + 4H^+ + 4e \rightarrow 2H_2O$$

and

$$O_2 + 2H^+ + 4e \rightarrow 2OH^-$$

Some OH⁻ ions drift over to the anodic side where they react with the Fe^{2+} ions produced during the anodic reaction:

$$Fe^{2+} + 2OH^- \rightarrow Fe(OH)_2$$

The rest of the OH⁻ ions react with the free hydrogen ions:

$$H^+ + OH^- \rightarrow H_2O$$

The ferrous oxide produced is then oxidised by free oxygen from the atmosphere or dissolved in the electrolyte to form hydrated ferric oxide (which is red):

$$4Fe(OH)_2 + O_2 \rightarrow 2Fe_2O_3.H_2O + 2H_2O$$

If the supply of oxygen is more restricted on the anode side, the final product is usually in the form of either the anhydrous or the hydrated magnetite, Fe_3O_4 or $Fe_3O_4 . nH_2O$

The rust formed has a porous structure and is quite loose. In consequence the reaction continues, the rate depending solely upon the magnitude of current passing between anode and cathode; this in turn depends upon the difference in potential between anode and cathode after taking into account the various polarisation effects and the resistance to the passage of the current as it passes from anode to cathode.

The most common form of galvanic corrosion occurs when steel components are joined to copper ones in a corrosive atmosphere. This is by no means the only case—others will be dealt with in Chapter 2. Whenever copper and iron are joined together, the iron closest to the copper corrodes most rapidly, because the path that the electrons must travel to enable the cathodic reaction to take place is the shortest.

1.17 Corrosion of a single metal

The corrosion of a single metal is caused by two effects (see Fig. 1.13):

1 The single metal is not really uniform at all, but consists of crystals having different chemical or physical characteristics. For example, in the case of iron and steel, the austenite/martensite phase consists of solid

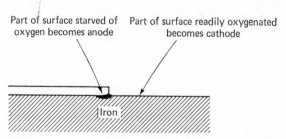

Part of surface starved of oxygen becomes anode   Part of surface readily oxygenated becomes cathode

Iron

Fig. 1.13. Corrosion of a uniform metal surface due to differential aeration

solutions of carbon in iron, the pearlite phase is virtually pure iron, and the cementite phase is the chemical compound Fe_3C. In addition, iron and steel castings or pressings contain small particles of pure carbon and other inclusions. All these have different $E°$ values. The crystals with the highest $E°$ values become the anodes, while those with the lowest $E°$ values become the cathodes. During the process of corrosion, polarisation effects can change the relative potential of neighbouring crystals. Crystals that have acted as cathodes and served to discharge the electrons and hydrogen ions produced can now become anodes and dissolve away, while crystals that previously acted as anodes may get lower $E°$ values due to various polarisation phenomena and change into cathodes. During a given process of corrosion the roles of the crystals may change many times over.

2 Different sections of the single metal are exposed to different oxidising conditions. This is mainly due to differences in the ease with which air can get to the surface of the metal (Fig. 1.13).

Even if the base metal has the same electrochemical potential throughout, there is a difference of electrical potential between the two parts of the surface which is equal to:

$$\Delta E = 0 \cdot 0148 \log_{10} \frac{a_1 (O_2)}{a_2 (O_2)} \quad V \qquad (1.16)$$

where a_1 and a_2 are the activities of the oxygen in contact with different sections of the base metal. As an approximation, a_1 and a_2 are represented by the partial pressures of the oxygen above the portions of the base metal concerned. The part of the metal surface where the activity a is greater then becomes the cathode of the corrosion cell, while the section which is starved of oxygen becomes the anode and dissolves away. The driving force of this corrosion cell is determined by the value ΔE as given in Eq. (1.16) and the rate of corrosion is once more governed by the current that passes from anode to cathode.

Literature sources and suggested further reading

1   UHLIG, H. H., *Corrosion and corrosion control*, John Wiley, New York (1963)

2   CHILTON, J. P., *Principles of metallic corrosion*, Royal Institute of Chemistry, London (1968)

3   GLASSTONE, S., *Textbook of physical chemistry*, Macmillan, London (1968)

4   SCULLY, J. C., *The fundamentals of corrosion*, Pergamon Press, Oxford (1966)

5   SPELLER, F. N., *Corrosion, causes and prevention*, McGraw-Hill, New York (1951)

6   CHANDLER, K. A., 'Prevention and control of corrosion', *Highways and Traffic Engineering* (September 1969)

7   MORGAN, L. H., *Cathodic protection*, Leonard Hill, London (1959)

8 BAKHALOV, G. T., and TURKOVSKAYA, A. V., *Corrosion and protection of metals*, Pergamon Press, Oxford (1965)
9 *Proceedings of the First International Congress on Metallic Corrosion*, Butterworths, London (1962)
10 *A background to the corrosion of steel and its prevention*, BISRA, London (1969)

Chapter 2   Galvanic and Differential Aeration Corrosion of Ferrous Metals

It is a popular misconception that the basic difference in electrochemical potential, as given in Table 1.2, is the real criterion regarding the rate of corrosion that actually takes place between a pair of metals.

One must not forget that this difference plus the difference in e.m.f. due to the concentration of ions is the reversible electrochemical potential only, i.e. the e.m.f. which applies when *no* current is flowing. From the point of view of actual corrosion the interest lies in the potential difference when a *given* current is flowing. A pair of metals, where one or both have good polarisation properties, is extremely stable against corrosion even if their reversible ΔE value happens to be large.

Unfortunately, the most important and useful of all metals—iron and steel—have poor polarisation properties. The corrosion product, rust, is porous, loose and flocculent and adheres badly to the base metal. This is why iron and steel corrode so badly, not because the $E°$ value is quite high at $+0.44$ V. This value is, nevertheless, lower than that of metals such as zinc, titanium, aluminium and magnesium, all of which are less liable to corrosion than iron because they show better polarisation characteristics.

2.1 Galvanic corrosion of steel in contact with metals that are more electronegative than steel

The most rapid form of corrosion that can take place with iron or steel is always caused when a galvanic couple is formed between the metal and an element which is more noble, such as copper, nickel, tin, lead, tungsten or antimony. Aluminium too can act as a cathode to increase the rate of corrosion of the iron or steel but for a different reason. Aluminium has some protective action at first for steel, as its e.m.f. is less. Once aluminium gets covered with an oxide film it becomes virtually an oxygen–oxide cell and causes rapid corrosion of the ferrous metals in contact with it.

In engineering design it is essential to avoid direct contact between iron or steel and one of the metals mentioned above if the joint is likely to be exposed to a corrosive atmosphere. However this is not always possible, but to minimise the rate of corrosion whenever it is necessary to join steel to metals that are cathodic to it, one or more of the following techniques may be used.

INSULATION OF THE JOINT

By insulating the metals from each other using plastic or rubber materials (Fig. 2.1), the flow of current from anode to cathode is impeded or even stopped, thereby slowing down corrosion. Such insulation has to be carried out very carefully, because even tiny corrosion currents can produce considerable corrosion. Electricity can also be carried by surface films of water.

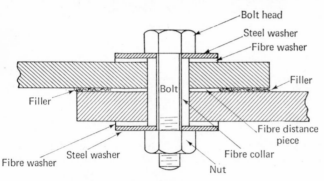

Fig. 2.1. Bolting two dissimilar metals together using electrical insulation to avoid galvanic corrosion

PAINTING

It is a mistake to paint the iron section and to leave the cathodic section unpainted. Paint films are always porous to some extent and may concentrate the anodic action to the restricted areas underneath the pinholes. The intense local corrosion that then takes place lifts off the paint film so that the rate of corrosion is increased to a very considerable value. The rate-determining part of the chemical reaction in the galvanic corrosion process is nearly always the cathodic reaction. A large unpainted area of copper or other cathodic metal ensures that this reaction proceeds very rapidly indeed. The best way to minimise corrosion is to paint the cathodic metal well, and to overlap on to the steel by about 10–20 mm. A band of steel, some 100–200 mm wide should then be left uncovered so that any galvanic corrosion induced is averaged out over a large area.

KEEPING CATHODIC AREAS SMALL

If it is necessary to join steel to a cathodic metal such as copper, the aim must always be to keep the cathodic surface areas as small as possible (Fig. 2.2). For example, steel nuts, bolts or rivets to hold large cathodic sections of metals such as copper, brass or oxidised aluminium together, are completely inadmissible. In such circumstances, the anodic reaction is concentrated upon a very small surface area and corrosion takes place extremely rapidly. The rate-determining reaction, which is the oxidation of the hydrogen ions and the elimination of electrons, proceeds on the large surface area of the cathodic metal. Conversely, the use of copper rivets to hold large steel plates together is perfectly all right. Because of the small area available for the cathodic reaction, the galvanic effect of the copper is quite negligible.

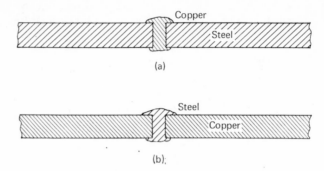

(a)

(b).

Fig. 2.2. The effect of anode and cathode areas on the rate of corrosion: *(a)* copper rivet in steel plate—corrosion insignificant; *(b)* steel rivet in copper plate—corrosion rapid

INTRODUCTION OF AN ADDITIONAL METAL

A third metal is used in close proximity to the joint to act as the anode. For example, if a joint between steel and copper is covered with zinc, the zinc becomes the anode instead of the iron. Because zinc exhibits much stronger anodic polarisation than iron, its rate of corrosion is quite low. On the other hand, its presence does prevent the iron from becoming the anode, and therefore corrosion is very much reduced.

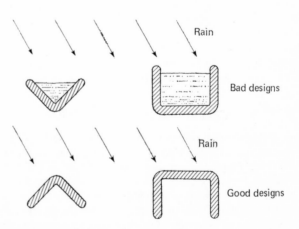

Fig. 2.3. Preventing rainwater from lodging on steel structures

ATTENTION TO JOINT DESIGN

Joints should be designed so that water or any other electrolyte cannot lodge near the joint (Fig. 2.3) and oxygen is denied access to the cathode portion of the corrosion cell. If, for example, steel pipe is joined to a copper pipe for the transportation of water, it is possible to avoid corrosion almost totally by deoxygenating the water. This is done by using a compound such as hydrazine, which reacts with dissolved oxygen. An alternative is mechanical deaeration.

2.2 Choice of metals that can be used in conjunction with iron

When considering which metals can be used safely together with iron and steel structures, it is necessary to examine whether the metal in question is clean and shiny, or whether it is covered with a thick and coherent oxide film. Provided the second metal has a reasonably unoxidised surface, steel can be used safely in conjunction with cadmium, zinc, aluminium, aluminium alloys, magnesium and the various magnesium alloys. Surface oxidation of the second metal considerably reduces the protecting power. In the case of aluminium and many of the aluminium alloys, the effect is, as has already been mentioned, the reverse to the one desired. Zinc, cadmium and magnesium, on the other hand, do not have such firmly adherent films as aluminium and because of this they maintain a strongly protecting action. This is particularly so with zinc, which is the best of all metals for protecting steel surfaces by galvanic means.

The galvanic action of pairs of metals is very much dependent upon the nature of the corrosive environment around the electrode pair. For example, while aluminium acts cathodic to iron and steel when it is in contact with steel in a normal, well oxygenated atmosphere which induces the formation of a very coherent film of Al_2O_3, it offers considerable protection to steel in a marine atmosphere. The reason is that aluminium chloride and sulphate are fairly porous salts, so that the natural anodic action of the aluminium comes into play, rather than its action as an oxygen electrode. Much the same applies to many of the magnesium alloys.

2.3 Distribution of attack at the joint between two metals

The nature of the attack on the joint between iron and a cathodic metal such as copper depends upon the conductivity of the electrolyte present. If the electrolyte is fairly clean water, corrosion is centralised at the joint itself, very close to the junction line (Fig. 2.4). Although the attack is slow, it is very deep. If the electrolyte contains dissolved salts, the rate of attack is much faster and also more even. It must be noted that at the narrow strip close to the joint, the rate of attack is no faster than if fairly non-conducting water only is present. Its magnitude is governed by the polarisa-

D

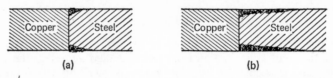

Fig. 2.4. The effect of the presence of salt on galvanic attack of copper-steel couples: *(a)* galvanic attack in salt-free water; *(b)* galvanic attack in water containing sodium chloride

tion and not by the size of the current. The rate of corrosion away from the joint is very much increased.

The maximum amount of attack is always caused by such ions as chloride and sulphate. This is the reason for the very high rate of corrosion in industrial atmospheres and in marine areas.

2.4 Corrosion of seemingly uniform steel sheet

When we look at a suitably etched steel surface under a microscope, we find that the so-called uniform surface is not uniform at all (Fig. 2.5). Steel

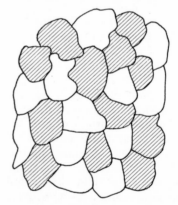

Fig. 2.5. In a sheet of steel anodic (hatched) and cathodic (open) crystals lie side by side

in fact, consists of a number of different solid phases—the most important of these are:

1 Martensite, which is basically a solid solution of carbon in iron.
2 Ferrite, basically pure iron.
3 Cementite (Fe_3C), a chemical compound of iron and carbon.
4 Particles of unreacted carbon.

The different pairs of steel components have different natural e.m.f's whereas carbon itself is very strongly cathodic. In the presence of an electrolyte, steel plate starts to corrode rapidly. Some of the crystals become anodes and the others become cathodes. At the anode, iron goes into solution and iron hydroxide is formed, being further oxidised by air to produce hydrated ferric oxide:

$$Fe \rightarrow Fe^{2+} + 2e$$
$$H_2O \rightarrow H^+ + OH^-$$
$$Fe^{2+} + 2OH^- \rightarrow Fe(OH)_2$$
$$4Fe(OH)_2 + O_2 \rightarrow 2Fe_2O_3 . H_2O \text{ (rust)}$$

The hydrogen ions released during the reaction are then oxidised to water at the cathode crystals with the help of atmospheric oxygen:

$$4H^+ + O_2 + 4e \rightarrow 2 H_2O$$

2.5 Galvanic corrosion due to working, heat treatment, etc.

A uniform piece of steel is often changed by drilling, cutting, etc., or by heating, into parts with different e.m.f's. This causes galvanic corrosion and is the reason why rusting is often localised near welds, holes, bends, etc., in finished steel. The answer to this is specially careful surface treatment at those positions.

If the water in contact with steel is either acidic, or contains salt, the reaction is very much accelerated, because of the increased ionisation of the electrolyte. In addition, corrosion products such as $FeCl_2$ and $FeSO_4$, are soluble in water and do not therefore cause as great a polarisation effect as the hydroxides and oxides of the ferric and ferrous iron.

Provided the pH is less than 3, hydrogen can be evolved directly and the cathodic reaction is no longer dependent upon the speed at which oxygen can diffuse at the cathode surface.

2.6 Change-over of anode and cathode surfaces

As ferric oxide has a much lower density than metallic iron it expands and spreads over to the cathode surface so that the cathode surface no longer has the electrochemical advantage of getting a better supply of air. The fact that the reaction.

$$Fe \rightarrow Fe^{2+} + 2e$$

has taken place on the anode side, causes concentration polarisation there. Therefore, even if the anode crystals started off by being the ones with a lower potential value, polarisation effects can reverse this so that, after corrosion has taken place for a short time, the potential of the cathode crystals is now the lower. In consequence the cathode crystals now start corroding, while previously anodic crystals become the cathodes. Because the rust is somewhat porous, it offers only some resistance to the supply of either electrolyte or air to the metal surface. Once corrosion is well under way, cathodic and anodic areas change places more or less continuously, with the result that the entire iron or steel sheet continues to rust through. The rusting is not regular and there will be pinholes and parts which go much deeper than the average. These locate positions where there was an excess proportion of electropositive metal crystals or where differential aeration conditions favoured particularly rapid corrosion.

2.7 Prevention of surface corrosion of sheet steel

Surface corrosion of sheet steel is prevented in three ways, preferably used in conjunction:

1 The use, in close proximity to the sheet, of a metal markedly anodic to steel, but which does not corrode rapidly itself because of polarisation properties. The most useful metal in this respect is zinc in the form of a direct coating. Magnesium coatings may also be used, but are far less effective.

2 Coating the metal by means of a film which protects it as far as possible from contact with air and water. Without the presence of air to oxidise the hydrogen ions at the cathode and an electrolyte to yield both positive and negative ions, corrosion becomes impossible. Later chapters will deal with a large number of such coating methods.

3 Design the object in such a way that there is no permanent contact of the steel surface with water. This is probably the most important method of protecting steel sheet against corrosion. Designs where water could be trapped without being able to evaporate readily are very prone to corrosion, even if the metal is well covered by a surface coating at those positions. The reason for this is that all paints and even plastic and rubber coatings are very slightly porous and cannot be relied upon to exclude water and air absolutely. Once corrosion starts, to even a slight extent, the lift-off of the surface coating material by the more voluminous rust soon increases the rate of corrosion.

Two British car designs of the 1950's were particularly prone to corrosion (Fig. 2.6). In one of these heavy corrosion took place on the wings just in front of the windscreen and in the other rust attacked just behind the headlights. The reason was the same in both cases. The design was such that damp soil collected just underneath the wings so that the steel sheet from which the wings were constructed was in permanent contact with moisture. In winter this also included a high proportion of salts used for the antifreeze treatment of roads. Other car designs suffered badly from corrosion at the bottom of doors. This was caused by rainwater running down the windows and thence to the bottom of the doors where it collected since no drainage holes were provided.

More modern cars, in the main, no longer suffer from these troubles. Wheel arches are nowadays designed so that they are perfectly smooth

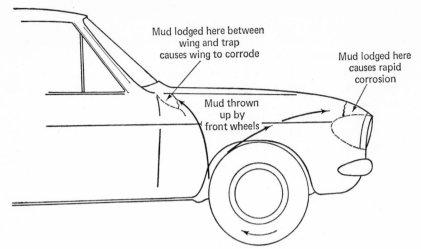

Mud lodged here between
wing and trap
causes wing to corrode

Mud lodged here
causes rapid
corrosion

Mud thrown
up by
front wheels

Fig. 2.6. The effect of 'mud traps' on the corrosion of motor-cars

inside and mud cannot lodge anywhere. The bottom edges of doors and
other parts where water can readily penetrate from the top have large holes
so that any moisture can readily seep away. In addition, many of the new
car designs use zinc coatings on parts which are likely to be in long-term
contact with moisture as well as sheathing with rubber and other surface
treatment materials.

The same considerations of design to prevent permanent contact with
moisture apply also to steel components used for other purposes. In the
opinion of the author, proper design is perhaps the most vital aspect of
corrosion resistance treatment:

STEEL OBJECTS NEED TO BE DESIGNED IN SUCH A WAY
THAT MOISTURE IS NEVER TRAPPED BUT CAN READILY
DRAIN OFF OR EVAPORATE.

2.8 Differential aeration corrosion

The differential aeration cell is a typical concentration cell, where the difference in potential between anode and cathode is caused by the difference in
the concentration of oxygen around them. This type of corrosion can afflict
steel sheet, usually in conjunction with the phenomenon of intergranular
corrosion previously mentioned.

2.9 Corrosion caused by drops of water on steel surface

When water, or worse still, drops of salt solution, etc., are spotted on
top of a steel plate, the part underneath the drop becomes very anodic

because it is starved of oxygen (Fig. 2.7). The area around the drop, being readily accessible to air, then becomes the cathode. This kind of corrosion is particularly likely under conditions of heavy dew, rather than persistent rainfall. The reason for this is two-fold:

1 Dewdrops are separate and therefore induce distinct anodic and cathodic areas, whereas with heavy rain the surface is wetted uniformly so that there is no distinction between the two areas.
2 Atmospheric impurities tend to be concentrated in dewdrops, thereby increasing the conductivity of the electrolyte. With rain, impurities are simply washed away.

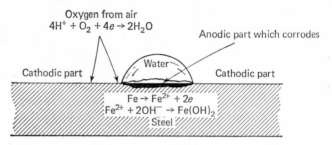

Fig. 2.7. Surface corrosion of steel plate underneath a water drop

Heavy corrosion of this type is common on cold metal surfaces subject to condensation, such as steel window frames, steel siding units in buildings, etc. Naturally this type of corrosion is best prevented by ensuring that condensation does not take place on the surface in question. This means that better insulation must be utilised so that the surface temperature of the steel object always remains above the dew-point of the atmosphere surrounding it. If this cannot be done, substantial surface treatment should be used. (It must be remembered that most paints are somewhat porous and therefore probably not as effective as thick plastic coatings.)

2.10 Crevice corrosion

By far the most common cause of differential aeration corrosion of steel objects is the existence of crevices formed when pieces of steel are joined together or present due to the specific design of the article. Rapid corrosion takes place at the crevice because the deep lying section is starved of oxygen and thus becomes the anode, while the surrounding area which is exposed to air, becomes the cathode. What makes matters worse is that crevices tend to get filled with moisture that cannot evaporate and have a small surface area, which concentrates corrosion damage.

Every attempt must be made at the design stage to avoid such crevices (Fig. 2.8). If crevices are unavoidable, they should be filled with a plastic mass or painted most carefully. Often crevices can readily be protected by anodic metals such as zinc.

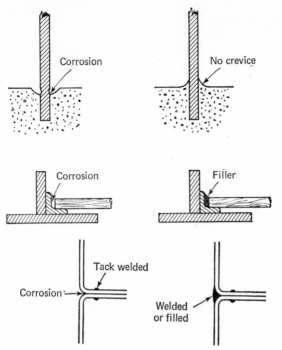

Fig. 2.8. Avoidance of crevices in design *(By courtesy of BISRA)*

2.11 Pitting due to impurities on the surface of steel

Corrosion takes place very often on surfaces where dust particles have settled (Fig. 2.9). This is particularly the case with small objects such as nuts and bolts. The reason once more is the difference in oxygen concentration over different parts of the surface and the presence of moisture. The part underneath the dirt particle is starved of oxygen and therefore becomes the anode, forming deep pits of corrosion. The surrounding part becomes the cathode and disposes of the electrons produced.

To avoid this type of corrosion, it is best to exclude dirty air from the surface of the metal. With large objects this is best done by using a protective sheet of polyethylene foil; small steel items are best kept in dustfree drawers.

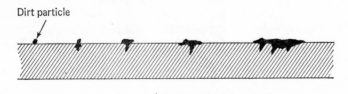

Fig. 2.9. The progress of pitting corrosion on sheet steel

2.12 Atmospheric corrosion of steel surfaces

Differential aeration, galvanic action, etc., induce the electrochemical potential difference which causes an electric current to flow. However, the rate of corrosion is mainly controlled by the rate at which the various chemical corrosion reactions can take place. One of the most important influences is the nature of the atmosphere in contact with the steel. In a very hot and dry atmosphere the rate of corrosion is very slow indeed. There is simply not enough electrolyte present. In very cold areas, such as Northern Canada and Northern Sweden, corrosion is low also. Apart from the fact that water vapour pressures are low, corrosion rates, like many other chemical reactions, are affected by temperature. Everything else being equal, metals corrode faster at higher temperatures that at lower ones.

The two atmospheric factors that cause the most rapid corrosion of iron and steel are:

1 A high chloride content of the air as in areas close to the sea.
2 A high sulphite–sulphate content of the air, together with a high concentration of carbon particles as in industrialised areas.

HIGH CHLORIDE CONTENT

In marine areas steel objects corrode rapidly due to the presence of sodium chloride in the atmosphere and in sea spray. An electrolyte containing sodium chloride readily conducts electric current and the chloride salts produced have a highly corrosive property as they are still soluble in water; this ensures that the steel at the anodic side is in constant contact with chloride ions. The rate of corrosion with a given concentration of salt in water is markedly temperature-dependent, being much faster in hot areas or during warm periods of the year than in cold regions. The rate of corrosion falls off very quickly as the distance from the sea increases. Even shielding metal from sea air has some effect in this respect.

An example is given by H. R. Ambler of the rate of corrosion of two similar steel plates, one located 50 m from the sea shore and the other at 3·2 km. The former corroded at a rate of 67 g/dm^2.year, while the latter corroded at a rate of only 3 g/dm^2.year. All this, of course, indicates the need for steel objects used in marine areas to be extremely well surface treated.

HIGH SULPHITE–SULPHATE CONTENT

Most of the atmospheric corrosion in industrialised areas is due to the presence of oxides of sulphur from the combustion of oil and coal. These materials also contain chlorides so that polluted industrial atmospheres have appreciable concentrations of hydrogen chloride. The following reactions apply:

$$S + O_2 \rightarrow SO_2$$
$$2SO_2 + O_2 + 2H_2O \rightarrow 2H_2SO_4$$
$$H_2SO_4 + 2NaCl \rightarrow Na_2SO_4 + 2HCl_{(gas)}$$

The combination of sulphuric acid and hydrochloric acid in the atmosphere has been found to have the most devastating corrosive attack of all on iron and steel—far greater than the effect of chlorides alone in a marine atmosphere or of sulphates alone. The rate of attack on ingot iron in a heavily industrialised atmosphere in Great Britain is between 0·08 and 0·10 mm/year, as against about 0·05 mm/year in a marine atmosphere. In the absence of either chlorides or sulphates, the rate of corrosion in similar humidity and temperature conditions is likely to be between 0·02 and 0·03 mm/year.

2.13 Effect of rust and impurities in the air

Although rust is by no means a coherent coating, it has some resistance to the passage of air and moisture. In consequence the rate of corrosion of rust-encrusted steel objects slows down markedly, provided the water in contact with the steel is reasonably free from sulphate, sulphite and chloride ions.

Appreciable corrosion cannot take place if the atmosphere is uncontaminated by impurities, provided the relative humidity always remains below 100%. In practice, however, due to temperature fluctuations, condensation on steel surfaces is likely to take place when temperatures drop. Many of the impurities present in the air in industrial atmospheres are hygroscopic. If the particles land on the surface of the steel they can remove water more easily from an atmosphere with higher humidity than from one with a lower humidity. The rate of corrosion in damp atmospheres is therefore greatly increased.

For this reason the rate of corrosion in polluted atmospheres is restricted by the relative humidity of the air. In clean atmospheres this is not so important, provided moisture can be prevented from condensing on metal surfaces.

2.14 Dust in atmospheres

Corrosion from dust in the air is considerable. It has a far greater effect than even the concentration of sulphur dioxide and other acid gases. In general, the dust content of the average city air is of the order of 2 mg/m³; really highly polluted atmospheres may reach up to 1 g/m³. The main components of air-borne dust are carbon, carbon compounds, metallic oxides, ammonium sulphate, salt particles and other mineral matter.

The reasons dust particles are of such importance in inducing corrosion are:

1 They can form galvanic cells with the sheet steel.
2 They can obscure part of the surface of the sheet steel and thus induce differential aeration corrosion.
3 They absorb moisture from the air because of their hygroscopic nature

and therefore form a close contact between electrolyte and metal surface.

4 They can absorb acid materials such as H_2SO_4 on their surfaces.

2.15 Corrosion or steel when submerged in water

The rate of corrosion of steel when in contact with even a highly polluted damp atmosphere is always far less than the rate of corrosion found when steel is in contact with water (Fig. 2.10). Corrosion is most rapid when the water covers only part of the object in question, e.g. when a container is partly filled with water. Corrosion then takes place near the interface between water and air. The section above the waterline is readily exposed to oxygen and thus becomes the cathode; the degree of corrosion is at a maximum just underneath the meniscus of the water and stretches downwards from it. Waterline corrosion can be stopped by the use of an inhibitor such as the phosphate, chromate, hydroxide or nitrite salts of sodium.

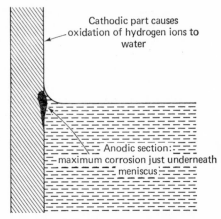

Fig. 2.10. Waterline corrosion

2.16 Differential oxidation corrosion in oil tanks, etc.

Corrosion of particular importance to the oil industry or other industries where organic liquids are being stored in steel vessels, is that which takes place at the bottom of tanks if traces of moisture have settled (Fig. 2.11). The bottom of the tank then becomes the anode and the reaction taking place is

$$H_2O \rightarrow H^+ + OH^-$$
$$Fe \rightarrow Fe^{2+} + 2e$$
$$Fe^{2+} + 2OH^- \rightarrow Fe(OH)_2$$

The area of the vessel in contact with the oil or other organic liquid becomes the cathode. Since this area is very large and is kept from cor-

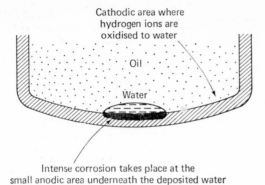

Fig. 2.11. Corrosion at the bottom of oil tanks

roding by the oil film on top, it acts as a most effective cathode. The oil has oxygen dissolved in it and, in general, the amount is sufficiently large to cause a rapid cathodic reaction:

$$4H^+ + 4e + O_2 \rightarrow 2H_2O$$

As oil often contains NaCl and other salts, reactions can be very rapid. Corrosion of this kind can be prevented by the following methods:

1 By constructing the bottom of oil-filled vessels with a metal that is cathodic to steel. Provided the outside of the vessel is duly protected against galvanic corrosion, all should be well.
2 By using a glass drainage vessel at the bottom of the vessel to enable any water to be drawn off.
3 By using an electric warning device embodying a circuit gap which becomes conducting when immersed in water or another aqueous solution. The current actuates an automatic drain valve to remove the water from the bottom of the vessel.

The same kind of corrosion is often found in oil pipelines and other equipment containing organic liquids if the design includes an elbow where water can collect. Great care must be taken in design to exclude such elbows wherever possible, as they are a sure source of trouble. If they cannot be avoided, provision must be made for draining off water accumulating there. The rate of corrosion that takes place at the bottom of oil filled vessels is unusually rapid because the anodic area is small and the cathodic area is large.

2.17 Flow differential oxidation corrosion

Corrosion can often take place in pipelines or other equipment in which aqueous liquids are being transported (Fig. 2.12). This usually occurs at positions where there is a difference in velocity beween different portions of

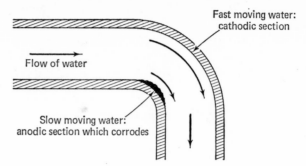

Fast moving water: cathodic section

Flow of water

Slow moving water: anodic section which corrodes

Fig. 2.12. Corrosion inside pipes carrying a flowing liquid

the liquid, i.e. at bends, nozzles, constrictions, etc. When liquid containing oxygen flows rapidly past a given section of pipe, the oxygen can be supplied far more quickly to the surface than it can in parts where the liquid is comparatively stagnant. In consequence the stagnant part of the pipe becomes the anode and corrodes; the section of the pipe in which the water moves rapidly becomes the cathode. The following reaction takes place there:

$$4H^+ + O_2 + 2e \rightarrow 2H_2O$$

This form of corrosion can be avoided only by ensuring that the water is properly deoxygenated. If this is impossible, sacrificial zinc anodes near the sections likely to be affected may seem to be the answer.

2.18 The effect of pH on the corrosion of iron and steel

In general, the rate of corrosion between pH 4 and pH 10 is virtually constant in normal aerated water, because in such a case the rate-determining factor is not the degree of ionisation of the electrolyte present, but the ease with which oxygen can pass to the cathode to oxidise the hydrogen ions to water (Fig. 2.13). Also, whatever the initial pH of the solution within this range, the surface of the iron is always covered with $Fe(OH)_2$ and free hydroxyl ions which give the cathode a pH of about 9·5.

When the pH falls below 4, other reactions start taking place. The ferrous oxide film dissolves and the lower the pH, the more rapid the rate at which the dissolution takes place. Also, the reaction

$$2H^+ + 2e \rightarrow H_2$$

is no longer stopped by considerations of polarisation and hydrogen starts to be liberated. Cementite (Fe_3C) has a lower hydrogen polarisation than purer forms of iron and therefore corrodes faster in an acid atmosphere than pure iron. Cold-rolled steel, which has a large quantity of cementite crystals, as a consequence corrodes fairly rapidly under the action of acids.

Concentrated oxidising acids such as nitric and sulphuric do not attack iron at all readily because they tend to form passive films thereby making it difficult for oxygen to reach the surface and get the cathodic action under way. (It is well known that concentrated sulphuric acid can be stored without trouble in mild steel drums.)

When the acids become diluted they attack iron very rapidly. The action of dilute nitric acid on iron is quite violent and oxides of nitrogen are produced. This is because the surface film produced is not impervious but porous, which allows the following cathodic reaction to take place:

$$2H^+ + 2HNO_3 + 2e \rightarrow 2H_2O + 2NO_2$$

When the pH of the electrolyte solution rises above 10, the pH of the cathodic surface of the iron, which is normally 9·5, rises to match that of the solution. Consequently, the polarisation effect increases and slows down the rate at which the cathodic reaction can take place; the higher the pH, the slower the rate of corrosion. For this reason it is general practice when running district heating pipelines to add a fairly large percentage of caustic soda to the circulating water. In such conditions, particularly if the water is also well deoxygenated by hydrazine, etc., corrosion of the mild steel pipelines is negligible.

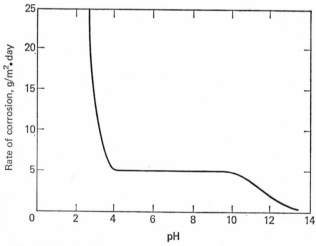

Fig. 2.13. Effect of pH on the rate of corrosion of steel in water

However, once the concentration of sodium hydroxide in water in contact with iron rises above about 10%, corrosion starts again as iron now begins to be dissolved as sodium hypoferrite, Na_2FeO_2, with the evolution of hydrogen:

$$Fe + 2NaOH \rightarrow Na_2FeO_2 + H_2$$

Iron in contact with a concentration of 43% NaOH corrodes at the rate of 0·10 mm/year at 20°C.

2.19 Influence of the presence of salts and air in water on the rate of corrosion of iron and steel

It has already been mentioned that the rate of corrosion of iron and steel is faster when in contact with water containing sodium chloride or other dissolved salts. This is because the conductivity of the water increases due to the increase in the number of ions present. On the other hand, the more concentrated the salt solution, the lower the solubility of oxygen in the water; consequently the rate of corrosion of steel in a salt solution is a maximum at a concentration of about 3% NaCl in water and falls off rapidly thereafter (Fig. 2.14). At a concentration of about 20% NaCl the rate of corrosion is about half that of the same sample of iron or steel in pure water.

Calcium and magnesium salts are markedly less corrosive than sodium salts with the same anions, e.g. calcium chloride is less corrosive than sodium chloride, since free calcium ions in solution react with carbondioxide to form impervious surface films of $CaCO_3$ on the metal surfaces; the equivalent sodium compound, Na_2CO_3, is soluble in water and therefore has no protecting power.

Town's water is purposely given an addition of some $Ca(HCO_3)_2$ to provide corrosion resistance to pipes and other vessels in contact with it.

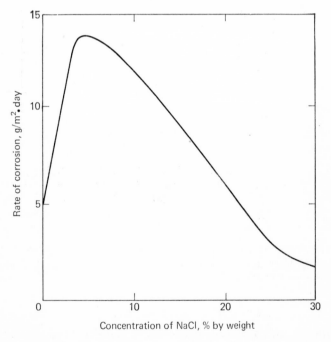

Fig. 2.14. Effect of sodium chloride concentration on the rate of corrosion of steel

Acid salts, when present in water, may cause corrosion due to combined hydrogen evolution and cathodic oxidation. They act like normal acids in this respect.

Alkaline salts, on the other hand, act as corrosion inhibitors. Typical salts of this type are

Trisodium phosphate	Na_3PO_4
Sodium tetraborate	$Na_2B_2O_7$
Sodium silicate	Na_2SiO_3
Sodium carbonate	Na_2CO_3

All these are widely used for this purpose.

Water mains made from cast iron are usually protected by adding some chromates or nitrites, but if the water is being heated and cooled intermittently, good effects are achieved by the use of 1·5% sodium benzoate and 0·1% sodium nitrite as inhibitors. The same materials are also effective in preventing corrosion by ethylene glycol solutions, after initial protection of the system by heating. Otherwise one needs to add about 5·0% sodium benzoate and 0·3% sodium nitrite as inhibitor to 25% ethylene glycol solution.

EFFECT OF OXIDISING SALTS

Oxidising salts can act in two ways:

1 Such salts as $FeCl_3$, $CuCl_2$, $HgCl_2$ and $NaOCl$ are not strong enough oxidising agents to make the iron or steel surface passive, but they are sufficiently strong to control the rate-determining cathodic action of oxidising the hydrogen ions evolved at the anode. These salts are extremely corrosive and speed up the rate of corrosion to a considerable extent.

2 Salts such as Na_2CrO_4, $NaNO_2$, $KMnO_4$ and K_2FeO_4 are very strong oxidising agents and make the steel surface passive. They slow down the corrosion reactions.

EFFECT OF DISSOLVED OXYGEN

The corrosion rate which affects iron or steel in contact with aerated water is practically linear, with respect to the concentration of oxygen (Fig. 2.15). When water at room temperature is completely saturated with oxygen, steel in contact with it can corrode at the rate of up to 10 g/m².day. In the absence of dissolved oxygen the rate of corrosion of iron and steel is virtually zero. For this reason one uses good deoxygenation practice when water, not intended for human consumption, is passed through a steel pipeline. The most usual methods of achieving this are:

1 The use of sodium sulphite
$$2Na_2SO_3 + O_2 \longrightarrow 2Na_2SO_4$$
2 The use of sodium sulphide
$$Na_2S + 2O_2 \longrightarrow Na_2SO_4$$

3 The use of hydrazine

$$N_2H_4 + O_2 \longrightarrow N_2 + 2H_2O$$

Deoxygenation of water intended for human consumption is usually carried out by mechanical deaeration using a centrifugal separator.

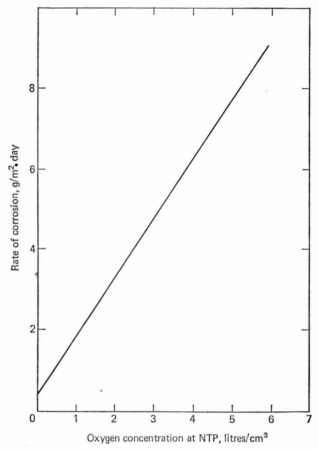

Fig. 2.15. Effect of oxygen concentration on the rate of corrosion in dilute salt solution

Literature sources and suggested further reading

1 UHLIG, H. H., *Corrosion and corrosion control*, John Wiley, New York (1963)
2 SPELLER, F. N., *Corrosion, causes and prevention*, McGraw-Hill, New York (1951)
3 *Corrosion prevention and design*, BISRA, London (1968)
4 *A background to the corrosion of steel and its prevention*, BISRA, London (1969)

5   CHANDLER, K. A., 'Preventing corrosion', *New Building* (October 1969)
6   CHANDLER, K. A., 'Prevention and control of corrosion', *Highways and Traffic Engineering* (September 1969)
7   STANNERS, J. F., 'A comparison of the corrosiveness of indoor atmospheres', *J. Appl. Chem.*, V 10, pp 461-470 (1960)
8   CHANDLER, K. A., and DREWETT, R., 'Corrosion of ferrous metals and its prevention', *IHVE Journal* (October 1968)
9   *The influence of salts in rusts on the corrosion of underlying steels*, BISRA, London (1968)
10 *Corrosion of buried metals*, BISRA, London (1952)
11 MORGAN, L. H., *Cathodic protection*, Leonard Hill, London (1959)
12 CHAMPION, F. A., *Corrosion testing procedures*, Chapman and Hall, London (1964)
13 BAKHALOV, G. T., and TURKOVSKAYA, A. V., *Corrosion and protection of metals*, Pergamon Press, Oxford (1965)
14 *Proceedings of the First International Congress on Metallic Corrosion*, Butterworths, London (1962)
15 GREATHOUSE, G. A., and WESSEL, C. J., *Deterioration of materials*, Reinhold, New York (1954)
16 BREGMAN, J. J., *Corrosion inhibitors*, Macmillan, London (1963)

E

Chapter 3   Miscellaneous Forms of Corrosion
That Affect Iron and Steel

3.1 Stray current corrosion

Buried pipelines and other objects often suffer severe corrosion from the effect of electric currents. If a conductor carrying direct current is run close to a steel pipe, current may leak to this pipe. This was a phenomenon that took place frequently with pipelines positioned close to tramlines but is now found mainly with pipelines buried close to the rails of underground railways, telephone cables, etc.

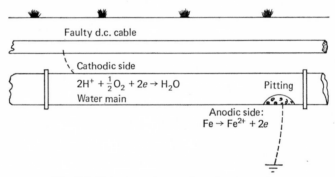

Fig. 3.1. Stray current corrosion of water mains

Trouble arises when the pipeline, etc., gives a possible path to the flow of electricity other than the original conductor (Fig. 3.1). What takes place in effect is an electrolytic process. At the anodic side of the buried conductor, iron is changed to ferrous ions, giving off electrons:

$$Fe \rightarrow Fe^{2+} + 2e$$
$$Fe^{2+} + 2OH^- \rightarrow Fe(OH)_2$$

On the cathodic side hydrogen ions are converted into water if oxygen is readily available:

$$2H^+ + \tfrac{1}{2}O_2 + 2e \rightarrow H_2O$$

In ordinary corrosion processes the e.m.f. between anode and cathode is not high enough to permit the liberation of hydrogen gas if there is no oxygen present. There is no need whatever for the presence of air to enable the cathodic reaction to take place where stray current corrosion is concerned because of the high e.m.f's involved. The cathodic reaction is simply

$$2H^+ + 2e \rightarrow H_2$$

The rate at which stray current corrosion takes place depends entirely on the current passing along the conductor. For every faraday of electricity passing (96,500 C) exactly 27·93 g of iron pass into solution.

Stray current corrosion always appears as pitting at certain positions on a pipe or other buried conductor. This is the position where the current drains away either to earth or to another conductor. Sometimes a corrosion cell appears in a buried pipeline and seems to be a reversal of the normal electrochemical series. For example, steel may corrode preferentially to aluminium; in such a case one can be certain that stray current corrosion is the cause.

3.2 Methods of preventing stray current corrosion

To paint an affected pipe for the prevention of stray current corrosion is usually worse than useless. All that happens is that the corrosion reaction is concentrated at pinholes and cracks in the paint and consequently the rate at which puncturing occurs increased. The main methods (see Fig. 3.2) used to prevent stray currents are the following:

1 Wherever possible alternating current should be used for underground cables. The amount of stray current corrosion induced by a.c. is less than 1% of that induced by d.c. The only exception concerns steel embedded in concrete that contains appreciable quantities of chlorides. Corrosion of the steel reinforcement in such conditions can be appreciable.

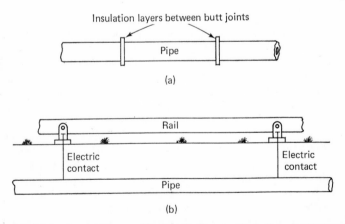

Fig. 3.2. Methods of protecting underground pipes against stray current corrosion: (a) insulated joints; (b) electric drainage

2 For gas pipes and oil pipes, a very satisfactory method of avoiding this kind of corrosion is to insulate the joints. This very much reduces the effect of stray current corrosion, except when the pipe is laid in very damp soil so that the soil bridges the insulation between pipe joints. This can cause the formation of a number of anodes and cathodes along each length of piping. Alternatively, each pipe section can be connected to a continuous conductor (a rail), so that there is an easier path for the current than through the pipe. The method of insulating the joints is not particularly useful for pipes carrying water or aqueous solutions because the current is then conducted from one section of pipe to the next by the contents of the pipe.

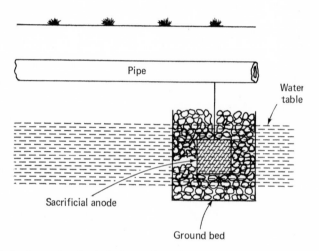

Fig. 3.3. Protection of a buried pipe using a sacrificial anode in a ground bed underneath the water table

3 A very successful system of protecting a pipeline is to connect a lump of pig iron, scrap steel, graphite, etc., to serve as the anodic part of the pipeline so that corrosion takes place on this so-called 'sacrificial anode' (see Fig. 3.3) and not on expensive pipe. These anodes are buried in a very corrosive environment, using coke-breeze surroundings or 'backfill'. Some of the most common sacrificial anodes used are given in Table 3.1

4 A widely used method to counteract stray current corrosion is to use electrical drainage. This means that the pipe or other buried object is connected to a source of power which has an e.m.f. that exactly counterbalances the e.m.f. induced by the stray current. This is sometimes difficult to achieve because the e.m.f.-causing stray currents tend to fluctuate. Many of the European district heating networks have been using systems of this type for many years, with electrical drainage at the expansion and inspection chambers of the pipeline systems.

TABLE 3.1

Material	Loss, kg/A.year	Maximum working capacity, A/m²
Pig iron	35	5
Scrap steel	45	5
Graphite	4	20
Silicon iron	4	40
Aluminium	20	20
Platinised titanium	Zero	10,000
Lead alloys (for sea-water only)	Almost zero	200
Magnesium	35	Depends on nature of backfill
Zinc	55	Depends on nature of backfill

3.3 Impressed currents

To protect systems by means of an impressed current it is necessary that the metal system should have the same surface potential throughout. This can be done by the use of a potentiostat which adjusts the potential applied to the e.m.f. differences of the system (Fig. 3.4). The best corrosion protection is applied when the potentials of metals are as follows: steel, -0.85 V; lead, -0.55 to -0.68 V; aluminium, -0.8 to -0.9 V. For large structures a transformer rectifier is used with the negative terminal

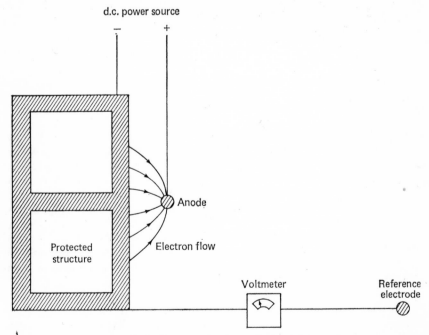

Fig. 3.4. Cathodic protection by an impressed current

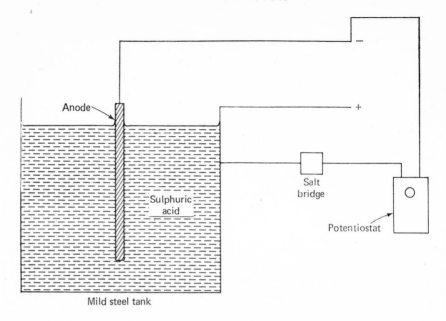

Fig. 3.5. Protection of an acid tank by an impressed current

connected to the protected structure and the positive one to a set of anodes (Fig. 3.5). The anodes can be made either of base metal, which will then corrode rapidly and must be replaced periodically, or of inert materials such as lead–platinum bi-electrodes, platinised titanium, copper, etc. The anodes should be spaced close together and at regular intervals (Fig. 3.6).

3.4 Stray current corrosion affecting ships

During the fitting-out period of a ship, powerlines are run from land to the ship and current induced corrosion is produced on the ship's hull wherever there is a crack or unevenness in the paint film. Stray current corrosion also affects ships in service due to the practice of using a ship's hull as the earth, thus making it anodic. This is prevented by the use of sacrificial anodes. In salt water conditions, magnesium is the most widely used—the magnesium becomes the anode which steadily dissolves, thereby making the rest of the ship's hull cathodic. Magnesium anodes are so designed that they last from refit to refit of the ship. Magnesium is unsuitable for fresh water as its surface then gets passive. Aluminium–calcium–zinc sacrificial anodes are used instead.

In some cases impressed power systems using graphite, platinum-clad silver and platinum-clad titanium as anodes have been used with great success. These are recessed into the hull of the vessel and serve to keep the hull cathodic by electrical drainage.

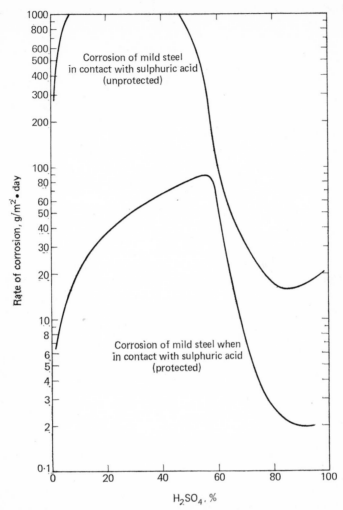

Fig. 3.6. The effect of impressed current on the corrosion of mild steel in contact with sulphuric acid

3.5 Corrosion induced by bacteria

As the name implies, anaerobic corrosion takes place in the absence of air and is, in fact, stopped by the ready access of oxygen. This type of corrosion particularly affects buried pipelines and other buried steel objects. The most common form of bacterially induced corrosion is the one which takes place in rich, acid clays found in industrial areas in northern England. The corrosion is caused by the action of micro-organisms about 2 μm in length and 10 nm in diameter. Its most common name is *Desulphovibrio desulphuricans* and it catalyses the reduction of sulphates to sulphides. The reactions that take place are

$$Fe \rightarrow Fe^{2+} + 2e$$
$$H_2O \rightarrow H^+ + OH^-$$
$$SO_4{}^{2-} + 8H^+ + 8e \rightarrow S^{2-} + 4H_2O$$

This is the reaction catalysed by the bacteria. It is somewhat exothermic, and the bacteria use the energy gained for their living processes. The S^{2-} ions then react with the Fe^{2+} ions to produce black FeS:

$$Fe^{2+} + S^{2-} \rightarrow FeS$$

Some of the ferrous ions, however, react with free hydroxyl ions to produce ferrous hydroxide.

Steel and iron suffer fairly rapidly from such bacterial action. Steel mains 10 mm thick have been punctured in as little as 9 years. This form of corrosion is prevented most effectively by making the soil alkaline, i.e. bringing the pH up to above 7·5. The presence of tannins in close proximity to the pipe also very much reduces the activity of sulphate-reducing bacteria.

The presence of oxygen close to the surface of pipes, i.e. embedding the pipes in a loose and sandy soil, has been found to eliminate bacterial corrosion.

OTHER CORROSION ACTIVITIES CAUSED BY BACTERIA

It has recently been shown that reduction of sulphates is not the only reaction catalysed by bacteria. Carbonates and nitrates are also liable to be reduced by hydrogen under the action of bacteria and thus provide a rapid cathodic action, which in its turn induces a more rapid anodic or corrosion action. The reactions are

$$CO_3{}^{2-} + 10H^+ + 8e \rightarrow CH_4 + 3H_2O$$
$$NO_3{}^- + 9H^+ + 8e \rightarrow NH_3 + 3H_2O$$

An important bacterial corrosion phenomena is that caused by the so-called 'iron bacteria' or *Gallionella ferruginea*. These bacteria catalyse the oxidation of ferrous salts into ferric salts and derive the energy needed for their living processes from this exothermic reaction. They rapidly oxidise soluble ferrous salts into a very voluminous form of ferric oxide, often obstructing the flow of water in pipes. Iron bacteria corrosion usually affects the inside of pipes only and is often coupled with corrosion caused by sulphate-reducing bacteria.

Corrosion by iron bacteria is prevented by providing a thick coating on the inside of the pipe. One of the best methods has been the use of a paint formulated with PVC resin and carrying aluminium flakes as a pigment.

PROTECTION OF PIPES AGAINST BACTERIAL ACTION

When coating pipes with continuous bandages that have been impregnated with bitumen or other materials, the use of materials such as hessian or cellulose fabrics must be avoided because these are liable to rot and their decomposition products provide just the kind of nutriment needed by the bacteria.

The undermentioned materials are suitable for the manufacture of water-proofing tapes:

1 Asbestos products (possibly not strong enough).
2 Woven glass cloth (expensive).
3 Plastics cloth which has been found to be resistant to rotting (most are still in experimental stage).

A typical specification for coating steel pipes that are to be used for transporting oil and gas in Holland is the one given below. It is intended to cater for really severe ground conditions. Four separate coating layers are prescribed:

1 A layer of blown asphalt bitumen, 0·5 mm thick.
2 A 5·5 mm thick coating of blown asphaltic bitumen carrying 30% in-organic filler. The fillers used are powdered talc, kieselguhr, slate, mica, limestone, pumice, etc.
3 A woollen cloth impregnated with blown asphaltic bitumen 2 mm thick, sandwiched between two layers of asbestos cloth or glass-fibre cloth.
4 A layer of blown asphaltic bitumen containing a high proportion of inorganic fillers, used as an external protective cover.

Although, for less severe conditions, it is possible to use less rigorous specifications, it is at all times essential to ensure that the external coating is highly robust and able to withstand rough handling. Also, it is necessary to ensure that there are no cracks or other imperfections in the coating of the pipe, as corrosion attack is then concentrated, causing early failure by puncturing.

Coating of pipes with either lead or zinc has not shown itself to be as satisfactory as the use of thick bitumen coatings. Lead is cathodic to steel in most soils, thereby accelerating corrosion rather than slowing it down. Zinc tends to dissolve in acid soils, but does give some short-lived protection. Coating pipes with zinc at the rate of 650 g/m² protects them in most soils for about 5 years.

The inside surfaces of pipes are treated with bituminous mixes carrying fillers; these are commonly applied by centrifugal processes. A method used to ensure that the coatings key firmly to the inside surface is to groove the surface during the manufacture of the pipe.

3.6 Intergranular and stress corrosion

All forms of steel consist of grains or granules (see Fig. 3.7). As already mentioned, the main constituents of these grains are:

1 Ferrite, which is pure iron.
2 Martensite, which is a solid solution of carbon in iron (cooled austenite).

3 Cementite, which is a compound with the formula Fe_3C.
4 Carbon (graphite flakes).

In normal circumstances the grains adhere extremely well and if a piece of steel is stressed beyond the elastic limit, crystals simply elongate by a process of sliding. This means that the crystals fracture and sections slide over each other to reform later on.

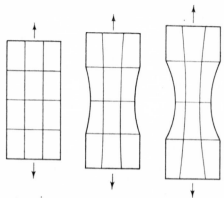

Extension of piece of steel under non-corrosive conditions
(schematic)

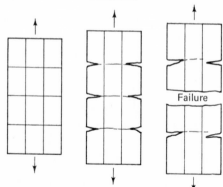

Failure

Extension of piece of steel under corrosive conditions
(schematic)

Fig. 3.7. Stress corrosion

If, however, the gaps between the crystals are opened up by corrosion, then stresses can produce more unfavourable phenomena. Stresses now act to enlarge the gap between the crystals and the metal literally tears in half, using the corroded hairline cracks as a starting tear. In consequence even extremely slight intergranular corrosion very much reduces the ultimate tensile strength of the steel component (see Fig. 3.8) and must be avoided at all costs.

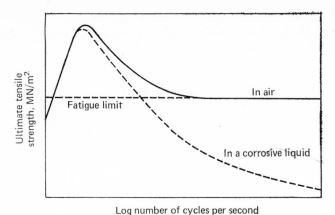

Fig. 3.8. Effect of corrosive environment on ultimate tensile strength of steel under the effect of vibration

Different forms of steel or non-ferrous alloys show markedly differing characteristics towards intergranular corrosion and therefore to stress cracking. The worst are those in which grain boundaries are highly anodic and therefore where corrosion is concentrated in small areas.

MOST COMMON FORMS OF STRESS CORROSION CRACKING OF IRON AND STEEL

Steel exposed to sodium hydroxide (caustic cracking)
Steam boilers, which were made by riveting construction, rather than being welded as they usually are today, are particularly liable to damage by caustic alkali solutions. But even modern boilers can suffer from this phenomenon unless careful water treatment (described in Chapter 10) is used. Crevices are formed at intergranular positions by corrosion processes, involving the solution of iron in concentrated sodium hydroxide (see Fig. 3.9), formed from normal alkaline boiler water by evaporation. The presence of silicates in the boiler water accelerates the formation of these minute intergranular hair cracks, which serve as 'starting tears' so that a sudden increase of steam pressure inside the boiler may cause failure of the boiler shell.

Steel exposed to nitrates
All nitrates tend to attack stressed steel by intergranular corrosion; this phenomenon is common in the chemical industry where hot nitrate solutions are being used. However, water at room temperature containing even small quantities of dissolved nitrates can harm stressed steel. In Portland, Oregon, USA, steel cables supporting a bridge failed after only 12 years' service because rainwater containing dissolved ammonium nitrate had accumulated in crevices formed by the anchoring of the cable to the base. In this case it was shown that the 0·7% C steel used for the construction of these cables was unusually subject to intergranular corrosion by ammonium nitrate.

Steel cables used externally must, therefore, be made from steel alloys that have been shown experimentally to resist attack by atmospheric nitrates which are always present in trace quantities in rainwater.

Steel exposed to liquid ammonia

A form of attack very common with many refrigeration plants is stress corrosion of steel in contact with liquid ammonia (anhydrous). The parts worst affected are the cold-formed heads and welds of steel tanks used to contain the liquefied gas. One can avoid this attack by heat treatment, which relieves internal stresses. The most usual way of reducing the liability of steel to attack by liquid ammonia is to add about 0·2% of water; this appears to act as an inhibitor to the liquid ammonia.

Steel exposed to traces of HCN

Corrosion cracking has been found with steel containers carrying gases which contain traces of hydrogen cyanide. On removal of the hydrogen cyanide gas the trouble disappeared.

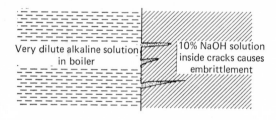

Very dilute alkaline solution in boiler — 10% NaOH solution inside cracks causes embrittlement

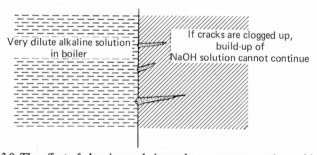

Very dilute alkaline solution in boiler — If cracks are clogged up, build-up of NaOH solution cannot continue

Fig. 3.9. The effect of clogging up hair cracks to prevent caustic cracking

PREVENTION OF STRESS CORROSION

The following methods are most commonly used.

Heat treatment

It has been found that heat treatment, by eliminating inherent stresses within the material, which are the ones likely to be attacked, effectively reduces liability to stress corrosion. The resistance obtained, however, is limited in time. An empirical relationship has been obtained:

$$\log_{10} t = 7,800/T - 9\cdot13$$

where t is the time in hours for which steel is resistant to stress corrosion and T the temperature in kelvins at which the steel has been tempered. From this can be evaluated how frequently retempering has to be carried out to ensure that the steel remains resistant to stress corrosion.

Shot blasting of surface
When metal surfaces are shot blasted, compressive stresses are formed at the surface; these tend to protect the intergranular parts against corrosion attack. Shot blasting therefore reduces the liability of steel to cracking but such surfaces remain effective only as long as they are continuous and have not degenerated by general corrosion processes.

Inhibitors used in solutions in contact with surfaces
Caustic cracking can be prevented by adding to the solutions materials that serve to seal the pores in the metal before they have been filled with sodium hydroxide in a harmful concentration. Such substances as quebracho and lignin are widely used, as well as waste sulphite liquors. Phosphates, sulphates and even sodium nitrates are used to prevent caustic cracking because they crystallise preferentially in any cracks and pores. However, sodium nitrate in neutral solution is itself a strong stress cracking agent.

Special stress-corrosion-resistant alloys
Very small quantities of metals such as aluminium, titanium, niobium and tantalum are added to steel because they react preferentially with carbon and nitrogen, thereby reducing the anodic character of the intergranular boundaries. Many of these alloys are highly resistant to stress corrosion.

3.7 Hydrogen cracking

During the pickling process when sulphuric acid is used to remove surface oxide films from steel, it is almost inevitable that some of the acid reacts with free metal:

$$Fe + H_2SO_4 \rightarrow FeSO_4 + H_2$$

In the case of high carbon steels and many non-stainless alloy steels the hydrogen evolved tends to penetrate the surface of the steel, making it brittle. What happens is that the hydrogen enters into solution with the iron, particularly when sulphur, selenium and arsenic are present in the steel. The hydrogen later splits off to form occluded hydrogen bubbles which have an immense internal pressure and thus cause internal stresses within grain boundaries.

Hydrogen embrittlement is often found with high tensile steel bolts which are pickled and cadmium-plated. Failure in such bolts is often delayed because it takes a certain time for the reactions leading to the formation of hydrogen bubbles within the sample to develop.

Hydrogen can also enter steel from certain furnace atmospheres at steel works and during welding operations. The latter is of particular importance and is one of the causes of 'weld decay'.

Finally hydrogen can enter steel if the latter is exposed to hydrogen sulphide, e.g. where water containing H_2S is in constant contact with alloy steels such as springs. The hydrogen sulphide breaks up chemically, the hydrogen is dissolved in the steel in the monatomic state and later is excluded in the form of high-pressure hydrogen bubbles from within the steel structure.

Experiments have been carried out to measure the actual pressure exerted by the interstitial hydrogen and figures as high as 10,000 bar have been obtained.

3.8 Prevention of hydrogen cracking

Hydrogen cracking is one of the main problems affecting such important industrial processes as the hydrogenation of fats, the manufacture of synthetic ammonia, etc., and can be solved only by improved metallurgical processes. Fine-grained steels and high chromium steels are reasonably hydrogen resistant. Another steel which is hydrogen resistant is one containing: 1·5% Cr, 0·5% Mo and 0·3% C and having a very low phosphorus and sulphur contents. Such steels should be used in conditions where exposure to hydrogen or hydrogen sulphide is inevitable.

Because of the liability of high tensile steel bolts and other high tensile components to be affected by hydrogen in pickling baths and similar situations it is advisable to avoid acid pickling altogether and use mechanical methods of oxide removal such as sand blasting.

Weld decay, due to the presence of hydrogen, is always a danger unless special weld decay resistant steels are used.

3.9 Radiation cracking

When steel is exposed to intense radiation in nuclear reactors or similar plant, lattice changes that are similar to those that occur when the metal is worked extremely hard, often take place. This means that metastable kinds of equilibria of steel—which are also usually the tough ones—are disturbed, giving rise to various types of phase discontinuities. Often these cause internal stresses which finally lead to cracking. A number of nuclear reactors in use at present have suffered from cracked stressed bolts, etc. The problem is one on which, up to the time of writing, insufficient information is available. Without doubt it will have to be solved by the development of special steels resistant to radiation cracking.

3.10 Corrosion fatigue

The difference between stress corrosion cracking and corrosion fatigue cracking is that the former takes place when there is a steady tensile stress in a corrosive environment, whereas the latter occurs under conditions of reversals of stress. If one starts with a piece of annealed metal, the ultimate tensile strength (UTS) normally increases when the piece of metal is subjected to vibrational stresses for a short time. The peak strength obtained is called the 'work-hardened UTS'. If the piece is then subjected to continuous vibrating stresses the UTS falls off steadily until it levels off at a given value which is called the 'fatigue limit'. With metals such as most steel alloys, titanium alloys and some others, the fatigue limit in dry air is very high and one can therefore consider that such metals are not unduly affected by fatigue conditions provided the surroundings are non-corrosive. Aluminium and copper alloys have very much lower fatigue limits, even in conditions of no corrosion.

When a piece of steel is subjected to vibrational stresses under corrosive conditions the UTS falls off much more rapidly with the number of reversals of stresses. In addition, no fatigue limit is normally formed (Fig. 3.8).

Table 3.2 gives the UTS of different metals after the samples have been vibrated 5×10^7 times at 25 Hz, under different conditions:

1. The sample is vibrated in well water containing: 200 ppm of Ca^{2+}, 18 ppm Mg^{2+}, 140 ppm Na^+, 200 ppm CO_3^{2-} and 175 ppm Cl^- by weight.
2. The sample is vibrated in river water containing: 1,500 ppm of Na^+, 200 ppm of Mg^{2+}, 500 ppm of SO_4^{2-} and 1,800 ppm of Cl^- by weight.

TABLE 3.2　How exposure to corrosive conditions reduces the fatigue strength of metals

Material	Fatigue limit in air, MN/m²	Fatigue strength of metal after 5×10^7 vibrations, MN/m²	
		(1)	(2)
0·16% C steel (quenched, drawn)	250	138	48
13·8% Cr, 0·1% C steel (quenched, drawn)	345	250	125
17% Cr, 8% Ni, 0·2% C steel (hot-rolled)	345	345	175
67·5% Ni, 29·5 % Cu, (annealed)	225	180	195
Copper (annealed)	65	68	68
98% Al, 1·2% Mn hard	70	38	26
Duralumin (tempered)	115	53	45

It can be seen from this table that copper and copper alloys have excellent resistance to corrosion fatigue; stainless steels are also very good in this respect. Normal carbon steels which do not contain appreciable additions of chromium are poor, as are aluminium and its alloys. It must be realised that these figures are only approximate.

3.11 Causes and cures of corrosion fatigue

Corrosion fatigue cracks in metal are usually transgranular and are often branched. The reason for such fatiguing is that under the cycling action deep seated corrosion cells are generated leading to surface pitting; the cracking continuing afterwards using the pits as starters. Corrosion fatigue cracks are commonly found in:

1. Shafts which are somewhat out of line and which are exposed to corrosive liquids.
2. Steel sucker rods in oil-wells. Such rods have a very short life because of exposure to oil-well brines.
3. Wire cables exposed to the weather.
4. Pipes carrying liquids that have widely fluctuating temperatures so that the pipes are subjected to cyclic expansion and contraction stresses.

Corrosion fatigue can be prevented or reduced by the following methods:

1. The use of inhibitors such as $Na_2Cr_2O_7$. Tap water containing 200 ppm of this substance has virtually no deleterious effect on 0.35% C steel.
2. Organic coatings which contain chromate pigments such as $ZnCrO_4$.
3. Metal coatings applied to steel electrolytically. The most usual are zinc and cadmium; but tin, lead, copper and silver have also been found to be effective.

3.12 Graphitisation

When molten pig iron, which contains a high percentage of carbon, is cooled rapidly, white cast iron is produced in which most of the carbon is held in the form of either a solid solution of carbon in iron or as the chemical compound cementite, Fe_3C. If molten pig iron is cooled slowly, the cast iron formed is a grey variety, which is a mixture of the ferrite type of iron together with crystals of graphite.

It is this grey cast iron which becomes liable to graphitisation if conditions are right. The graphite particles act as very strong cathodes against the iron, the e.m.f. applied being as high as 2 V. The iron therefore goes into solution with the graphite remaining intact. As the total surface area of iron shrinks due to corrosion, the anodic processes are concentrated on the quantities of iron remaining. In due course the entire surface of the cast iron object becomes graphite and the change then continues to take place in depth. The phenomenon is particularly bad with underground cast iron pipes that have been in the soil for a long time. Instances may be found where virtually the whole iron content had been leached out in the form of iron salts, leaving only the graphite behind. This is often not detected until it is noticed that such pipes crack under the slightest distortion because the strength of the graphite is naturally much lower than that of the original cast iron pipe. White cast iron, which does not have free graphite present is virtually immune to graphitisation. Grey cast iron is not generally considered to be a suitable material to be kept under constant corrosive conditions because the presence of highly cathodic flakes of graphite can induce the virtual removal of iron from the object. It is far better to use mild steel instead as the amount of carbon present is very small.

3.13 Growth of cast iron

When cast iron is alternately heated and cooled it is possible for volume changes to take place which affect the accuracy of machined surfaces (Fig. 3.10). This phenomenon can affect brake drums, cast iron cylinder heads,

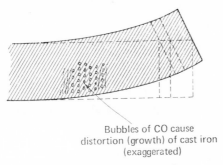

Bubbles of CO cause
distortion (growth) of cast iron
(exaggerated)

Fig. 3.10. The growth of cast iron

etc. The cause of such changes of volume, or 'growth', as it is called, is the transformation of the alpha phase of the iron into the gamma phase. During this change pores are opened up enabling some of the graphite particles to be oxidised to carbon monoxide. These small bubbles of gas remain occluded and induce internal stresses that cause the object to warp. It has been found that temperatures around 650°C are the worst for the growth of cast iron which is only found in the presence of oxidising atmospheres. The phenomenon can be prevented by a number of specialised metallurgical methods. If large quantities of silicon are added the transformation temperature is raised considerably and little growth takes place below about 800°C. Nickel, chromium and copper can be added to the cast iron to reduce the rate of oxidation of the graphite particles.

Both graphitisation and growth of cast iron can be reduced considerably by the use of more modern methods of making cast iron in which large particles of graphite are absent but where instead, the graphite is present only in the form of tiny granules.

F

3.14 Corrosion of steel embedded in concrete

Steel is used in conjunction with concrete in a wide variety of reinforced, prestressed and post-tensioned forms. Clean steel, embedded in Portland cement, is normally unattacked for many years due to the inhibiting effect of free lime present. In the absence of chlorides and sulphates, small patches of rust on the reinforcement steel merely serve to increase the strength with which the concrete is held to the steel. But if chlorides and sulphates are present in the concrete mix, really rapid corrosion may take place at isolated places on the steel reinforcement with the production of pockets of rust that can then crack the concrete. Calcium chloride should never be used as an antifreeze in prestressed or post-tensioned constructions as the risk of corrosion failure is very high. Many authorities also do not recommend the use of calcium chloride for simple reinforced structures, although it is supposed to be less harmful than sodium chloride.

Nitrates present in concrete have little over-all corrosion affect on embedded steel but must be avoided if prestressing or post-tensioning is considered because they induce stress corrosion, i.e. the sudden failure of the steel strand.

Provided the concrete is reasonably dense, corrosion on embedded steel strands and rods is slight only, because the cathodic (oxidation) reactions are normally impeded. Corrosion has, however, taken place in reinforced aerated concrete and similar materials so that special precautions must be taken. Concrete formulations that seek to avoid, as far as possible, the presence of chlorides and sulphates should be used when steel is embedded.

3.15 Fretting corrosion

This is a form of corrosion caused by wear (see Fig. 3.11). It is most commonly observed when there are two closely fitting parts that vibrate against each other. A typical example of such corrosion can be found in ball-bearing races that are pressed on to the shaft. The contacting surfaces can become badly pitted. Fretting corrosion often causes failure of suspension springs, belt heads, electrical contacts and bearings. What happens

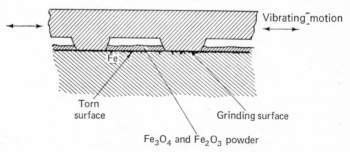

Fig. 3.11. The mechanism of fretting corrosion

is that the surface of the steel becomes torn and the particles then oxidise to form Fe_2O_3 and Fe_3O_4 in the presence of air and under the action of the heat produced by friction. Fretting corrosion is at a minimum in the absence of air or oxygen; therefore a method of preventing it is to use a protective atmosphere of nitrogen and hydrogen around the wearing surfaces.

In the absence of air wear can still take place but the debris is then only iron dust instead of iron oxide. It has been found that the rate of fretting in the absence of air is as low as 15% of that which takes place when oxidation reaction can occur.

It has also been found that fretting damage is less when there is 50% relative humidity than if the relative humidity is either 0% or 100%. At 50% relative humidity soft iron hydroxide, which has a lubricating action, is then formed. Frettage has often been a major cause of fatigue because the kind of damage produced causes cracks to start. To prevent fretting corrosion, it is best to coat surfaces with other metals, which act as lubricants. Copper, tin, silver, cadmium and even gold are applied in minutely thin layers on surfaces liable to fretting. Thin plastic or rubber films are also widely used to lubricate surfaces, the most popular being PTFE (polytetrafluoroethylene). Oil lubrication reduces friction and also prevents oxidation by keeping air away from any abraded particles.

3.16 Cavitation damage

Cavitation can occur wherever irregular water flow takes place, particularly if high pressures and vacua are involved. Immense sudden pressures are produced when vacuum bubbles collapse; these have been assessed as high as 150–200 MN/m^2. This causes formation of cavities in the steel and damage can be very rapid indeed. For example, a cast iron pipe 12·7 mm thick in a diesel engine can be perforated by cavitation in some 300 hr.

Cavitation damage is partly mechanical and partly chemical. The mechanical damage permits easy access for corrosive liquids which then continue to destroy the metal by standard corrosion reactions.

The prevention of cavitation is best carried out by designing equipment that carries fluids at speed in such a way as to prevent radical pressure differences in the system. If these cannot be avoided cavitation damage can be reduced by ensuring that the parts most affected by 'water hammer' are also the most cathodic, so that the oxidation reaction that causes most of the damage is avoided. Normal sacrificial anode methods using zinc or magnesium may be used.

3.17 Chemical attack on iron and steel

Mild steel has considerable resistance to strong alkalis, most neutral substances such as solvents, as well as very strong acids (see Fig. 3.12). Sulphuric acid with a concentration of more than 65% H_2SO_4, mixtures of

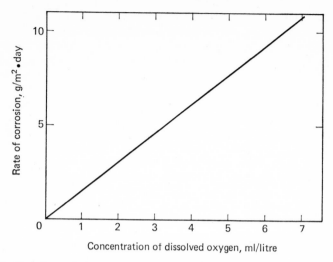

Fig. 3.12. Variation of the rate of corrosion with dissolved oxygen concentration at 25°C

concentrated sulphuric and nitric acids as well as anhydrous hydrofluoric acid with more than 63% HF have little effect on steel containers at room temperature. Hydrochloric acid attacks steel readily when hydrated reaching a maximum at a concentration of 33% HCl. As the concentration increases the corrosion rate falls rapidly. Anhydrous HCl has virtually no effect upon steel at all. Acetic acid and other organic acids that undergo negligible ionisation have little corrosion effect on mild steel when concentrated or dilute, except when dissolved oxygen is also present (see Fig. 3.13). Nitric acid dissolves mild steel or iron at room temperature although the degree of attack is very much reduced at concentrations above 40% as the iron is

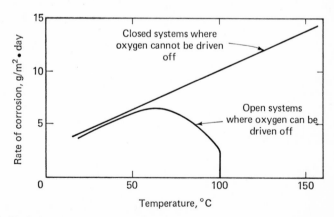

Fig. 3.13. Variation of the rate of corrosion at different temperatures with open and closed systems

then made passive. The reason why dilute nitric acid has such a strong corrosive effect is that it acts as its own oxidiser at the cathodic side of the iron.

The corrosion resistance of steels to acids is markedly increased when they are alloyed with substantial percentages of silicon (see Fig. 3.14). One alloy, Durion, containing between 14 and 15% silicon, has in contrast to normal mild steel, considerable chemical resistance to nitric acid.

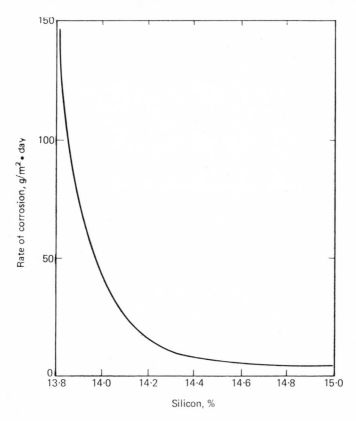

Fig. 3.14. Effect of addition of silicon on corrosion resistance of mild steel to 10% H_2SO_4 at 80°C

3.18 Corrosion of metals by acid vapours from timber

It has been observed for many years that iron and steel or other metals in contact with wood tend to corrode more quickly than the same metals on their own. This is mainly due to the emission of acetic acid vapours from the wood. It has been found experimentally that kiln drying of wood very much increases the rate of corrosion of metal objects in contact with the timber. It is suggested that during kiln drying breakdown of certain of the timber constituents to acetic acid can take place and this affects the degree of corrosion of metal objects in contact with such wood. Table 3.3 indicates the rate of corrosion by *vapours* from a 1% acetic acid solution and that caused by pure water vapour—both over a period of six days at 30°C with the metal exposed at 100% relative humidity. The figures give an indication of the kind of corrosion to be expected for different metals when they are placed in contact with timber without adequate ventilation.

TABLE 3.3

Metal	1% acetic acid solution, g/m²	Distilled water, g/m²
Steel (with 0·1% C)	328	2
Aluminium (99%)	23	3
Zinc (99·95%)	131	53
Cadmium (99·5%)	99	60
Magnesium alloy (95·5% Mg)	20	1
Brass (70/30)	12	Nil
Copper	22	Nil
Nickel	9	Nil
Tin	Nil	Nil

Literature sources and suggested further reading

1 BRESLE, A., *Recent advances in stress corrosion*, Swedish Academy of Science, Stockholm (1961)

2 BREGMAN, J. J., *Corrosion inhibitors*, Macmillan, London (1963)

3 UHLIG, H. H., *Corrosion and corrosion control*, John Wiley, New York (1963)

4 SPELLER, F. N., *Corrosion, causes and prevention*, McGraw-Hill, New York (1951)

5 *Corrosion prevention and design*, BISRA, London (1968)

6 *A background to the corrosion of steel and its prevention*, BISRA, London (1968)

7 *Prevention of corrosion of steel in supply waters*, BISRA, London (1969)

8 GREATHOUSE, G. A., and WESSEL, C. J., *Deterioration of materials*, Reinhold, New York (1954)

9 *Corrosion of buried metals*, BISRA, London (1952)

10 MORGAN, L. H., *Cathodic protection*, Leonard Hill, London (1959)

11 MADAYAG, A. F., *Metal fatigue. Theory and design*, John Wiley, New York (1969)

12 MANN, J. Y., *Fatigue of materials*, Melbourne University Press, Melbourne (1967)

13 PUTILOVA, I. N., BALEZIN, S. A., and BARANNIK, V. P., *Metallic corrosion inhibitors*, Pergamon Press, Oxford (1960)

14 CHAMPION, F. A., *Corrosion testing procedures*, Chapman and Hall, London (1964)

15 BAKHALOV, G. T., and TURKOVSKAYA, A. V., *Corrosion and protection of metals*, Pergamon Press, Oxford (1965)

16 *Proceedings of the First International Congress on Metallic Corrosion*, Butterworths, London (1962)

17 CHANDLER, K. A., 'Preventing corrosion', *New Building* (October 1969)

18 CHANDLER, K. A., 'Prevention and control of corrosion', *Highways and Traffic Engineering* (September 1969)

19 STANNERS, J. F., 'A comparison of the corrosiveness of indoor atmospheres', *J. Apl. Chem.*, V 10, pp 461–470 (1960)

20 CHANDLER, K. A., and KILCULLEN, M. B., 'Survey of corrosion and atmosperic pollution in and around Sheffield', *British Corrosion Journal*, V3(2), (1968)

21 CHANDLER, K. A., and DREWETT, R., 'Corrosion of ferrous metals and its prevention', *IHVE Journal* (October 1968)

22 *The influence of salts in rusts on the corrosion of the underlying steel*, BISRA, London (1968)

23 CLARKE, S. G., and LONGHURST, E. E., 'The corrosion of metals by acid vapours from wood', *J. Appl. Chem.* (November 1961)

24 GROMOBOY, T. S., and FRAUNHOFER, J. A., 'Corrosion prevention by impressed currents', *Chem. & Proc. Eng.* (April 1969)

Chapter 4　Corrosion of Non-ferrous Metals—I:

Aluminium, Beryllium, Chromium, Cobalt, Copper, Gold, Iridium, Lead, Magnesium

4.1 Aluminium

Aluminium has a density of 2·7 kg/dm³ and a melting-point of 659°C. It has good corrosion resistance even in polluted atmospheres and in many aqueous media despite the fact that its standard oxidation potential

$$Al \rightarrow Al^{3+} + 3e \qquad E° = +1\cdot66 \text{ V}$$

at 25°C is about 1·2 V above that of iron. The reason for this is that the active aluminium surface readily becomes passive on exposure to water. The passive film on the surface of aluminium consists of Al_2O_3 and is between 2 and 10 μm thick.

High-purity aluminium has the best resistance to corrosion. Its liability to corrode is markedly increased when other metals are in contact with it since virtually all other common metals, with the exception of magnesium, are strongly cathodic to aluminium, thereby increasing its rate of corrosion. Alloys, although stronger than pure aluminium, are more liable to corrosion than the pure metal. Aluminium and its alloys are more vulnerable than iron and steel to crevice corrosion and hidden joints must be studiously avoided in design.

ATMOSPHERIC CORROSION

Aluminium surfaces weather by a characteristic process of pitting, which falls off virtually to zero after the first year or so. The rate of corrosion is very much less than with steel and many other materials. In normal circumstances the rate of corrosion of aluminium roofing sheets during the first year varies between 0·01 and 0·02 mm even in fairly polluted atmospheres. Nor is there any appreciable loss of strength when aluminium or its alloys are exposed to corrosive atmospheres for long periods. Experiments were carried out on 1 mm thick aluminium alloy NS3 sheeting and it was found that its strength was reduced by 2% after one year's and by 3% after ten years' exposure to a marine atmosphere.

Aluminium alloys which have high copper content and are subsequently heat-treated can become subject to intercrystalline attack when exposed to corrosive atmospheres. Under such conditions the strength of the alloy

66

is reduced sharply on prolonged exposure. Such aluminium alloys must always be well painted when used in marine or industrial atmospheres. They are:

BS1490 (cast): LM 1-M, 3-M; LM 11-W and WP; LM 12-WP
BS1470–77 (wrought): H14-T; H15-W; H15-WP.

CORROSION IN AQUEOUS LIQUIDS

The rate of corrosion of aluminium at room temperature is high below pH 5·5 and above pH 8·5 (see Fig. 4.1). As the temperature of the solution is increased the aluminium becomes more sensitive to even slight alkalinity, although its sensitivity to acidic conditions is reduced. In boiling aqueous liquids aluminium is least attacked within the pH range 4·5 to 7.

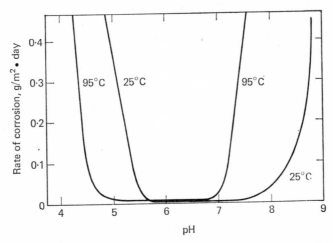

Fig. 4.1. Rate of Corrosion of aluminium under different pH conditions

Unlike iron and steel, corrosion of aluminium and its alloys is extremely heavy in alkaline solutions because of the ready formation of the complex ion AlO_2^-.

The full reaction which takes place between aluminium and sodium hydroxide is

$$2Al + 2NaOH + 2H_2O \rightarrow 2NaAlO_2 + 3H_2$$

Such a reaction takes place readily at room temperature; for this reason one must *never* use aluminium in contact with even very dilute alkaline solutions.

When aluminium is in contact with dilute acids very little happens at first provided the aluminium is passive, i.e. it has a reasonably thick oxide film. The period of induction is much longer with high-purity aluminium than with aluminium alloys. With pure aluminium, induction periods of several days are not uncommon, but they may be only a matter of minutes with very impure samples. After sections of the oxide film have been dissolved, the rest of the oxide acts as a cathode. The parts where the passivity has been removed now act as anodes, dissolve in the acid and produce hydrogen. The speed of reaction is limited by the conductivity of the remaining cathode film.

Normal pure aluminium dissolves slowly in dilute mineral acids, but electropolished aluminium which has only a minutely thin oxide film reacts almost violently with fairly concentrated hydrochloric acid. Aluminium alloys that contain copper, even in minute traces, also react vigorously with acids.

Apart from acidity and alkalinity, aluminium is also subject to attack by chloride ions, particularly in crevices and other hidden places, because under such conditions passivity may break down due to a differential aeration effect. This effect is enhanced when there are the merest traces of Cu^{2+} or Fe^{3+} contamination. What happens is that a galvanic cell is produced between the deposited copper or iron and the aluminium.

Aluminium is, in general, a satisfactory metal for use with soft water or deionised water from which all heavy metals have been removed. It should *not* be used for pipes carrying normal drinking water or industrial water; such waters usually contain traces of heavy metals, which can cause corrosion. Minute traces of Cu^{2+} ions are particularly harmful.

Aluminium salts are non-poisonous and therefore vessels made from aluminium are satisfactory for handling foodstuffs.

CORROSION OF ALUMINIUM BY OTHER MATERIALS

Mercury and mercury salts

Mercury and mercury salts attack aluminium extremely rapidly, since when they are brought into contact with an aluminium surface an aluminium amalgam is formed. The aluminium metal then diffuses through the amalgam film and comes into contact with atmospheric oxygen at the surface, without the benefit of a protective film. The highly exothermic reaction

$$4Al + 3O_2 \rightarrow 2Al_2O_3$$

takes place and a loose film of aluminium oxide is formed; this has no protective action whatsoever. Further quantities of aluminium diffuse through the amalgam film to be oxidised in their turn, a process which takes place rapidly due to the fact that the sample has now become quite hot, which accelerates the rate of reaction.

Lime, mortar and cements

Fresh slaked lime $Ca(OH)_2$ attacks aluminium with the evolution of hydrogen. But the corrosion products formed, which include fairly insoluble calcium aluminate, protect the rest of the surface of the aluminium against further attack. The reaction of aluminium with hardening or hardened Portland cement is quite slow. Aluminium alloys with an appreciable copper content should be avoided in conjunction with building materials.

Chlorides, amines, etc.

Apart from the liablity of aluminium to be attacked by sodium chloride and calcium chloride solutions, the material is also attacked readily by all chlorinated organic solvents, e.g. CCl_4, $CHCl_3$, $C_2H_4Cl_2$, etc. These react violently with aluminium. Contact of aluminium equipment with any such chlorinated hydrocarbons must be avoided at all costs because reactions could be quite dangerous.

The reaction between aluminium and carbon tetrachloride is

$$2Al + 6CCl_4 \rightarrow 3C_2Cl_6 + 2AlCl_3$$

This reaction is highly exothermic and the rate of solution of 99·99% aluminium in anhydrous CCl_4 is as high as 38 g/dm^2.day. It can happen that due to the large quantities of heat evolved, some aluminium may approach its melting-point and under such conditions the reaction proceeds explosively. Other alkyl chlorides and bromides also attack aluminium. Aluminium is readily acted upon by some of the strongly alkaline organic amines, but not by ammonia. Anhydrous alcohols attack aluminium at elevated temperatures but the reaction is inhibited by traces of water.

Inorganic acids

Aluminium is attacked by strong inorganic acids such as HCl, H_2SO_4, HF and H_3PO_4. It also reacts with some of the stronger organic acids such as formic, trichloroacetic and oxalic acids.

Hydrochloric acid is particularly harmful because it causes intergranular attack which very much weakens the structure of the metal and makes it liable to stress cracking.

The rate of attack of nitric acid at room temperature is a maximum at 20% concentration (see Fig 4.2). As the concentration increases the rate falls virtually to zero at 80% concentrations and above: the reason is probably the formation of substantial oxide films on the surface of the aluminium. The rate of corrosion is approximately doubled as the temperature increases by 10°C. Impurities such as traces of sulphuric acid and oxides of nitrogen also speed up the rate of corrosion.

GALVANIC COUPLING OF ALUMINIUM

Although open and exposed surfaces of aluminium are many times more resistant to corrosion by either aqueous liquids or polluted air than equivalent sheets of steel, aluminium is far more sensitive than steel to galvanic action. If aluminium is coupled to copper or a copper alloy, very rapid corrosion of the aluminium takes place at the joints.

Zinc is usually anodic to aluminium in either acid or neutral solutions and has therefore some protecting power. It is cathodic in alkaline solutions and thus accelerates the rate of corrosion. In rural areas steel can be coupled to aluminium, but in marine areas the rate of attack on aluminium is increased when it is joined to steel. Magnesium and aluminium can be coupled together without either metal being adversely affected by corrosion.

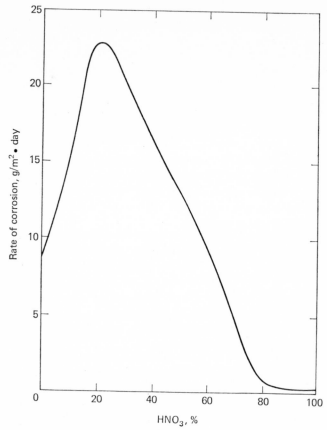

Fig. 4.2. Rate of corrosion of aluminium in HNO₃ at 25°C

The metal closest to aluminium in its potential in normal environment is cadmium. This is why cadmium-plated steel screws, bolts, clips, etc., are by far the best to use as fasteners for aluminium sheeting and other aluminium components.

ALUMINIUM ALLOYS

Aluminium is commonly alloyed with one or more of the following elements: copper, silicon, magnesium, zinc and manganese. Manganese alloys have as good or perhaps even better corrosion-resistance properties than pure aluminium.

The duralumin types contain copper and this is found in the form of tough $CuAl_2$ particles at grain boundaries. Duralumin is liable to stress corrosion cracking; pure aluminium is more or less immune to this. For this reason duralumin is often sandwiched between sheets of pure aluminium to make use of the good strength characteristics of the duralumin, yet avoid the troubles of stress cracking. The aircraft industry makes extensive use of 'clad sheets' of this type.

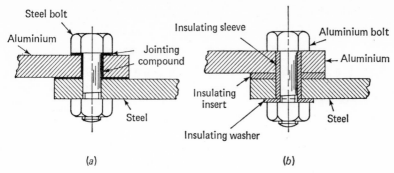

Fig. 4.3. Use of (a) jointing compound and (b) insulating washer, insert and sleeve at bolted joint of dissimilar metals *(By courtesy of The Aluminium Federation)*

Stressed or cold-worked 5–7% magnesium–aluminium alloys are also liable to stress corrosion cracking in damp air. Again the damage can be avoided with such alloys by cladding sheets with pure aluminium, which is not affected by stress corrosion.

RESISTANCE TO CHEMICALS

Aluminium has excellent resistance to the following chemicals:

1 Most organic acids, such as fatty acids, and reasonably concentrated acetic acid.
2 Nitric acid with a concentration in excess of 80% provided the temperature is kept below about 50°C.
3 Ammonia. Any concentration, both hot and cold.
4 Sulphur, hydrogen sulphide and mercaptans.
5 Fluorinated refrigerant gases such as the 'Freons'.
6 Atmospheric exposure to even highly polluted atmospheres. (Resistance to marine atmospheres is fair.)
7 Distilled water.

4.2 Beryllium

Beryllium is a very light metal with a density of only 1·8 kg/dm³. It has a melting-point of 1,530°C. It is nowadays becoming far more widely used because of the expansion in the development of nuclear reactors of all types. Beryllium has the useful properties of being able to moderate (slow down) neutrons and of having very low neutron absorption.

The main alloy of beryllium used is one containing 74% beryllium and 26% lithium.

Beryllium has good resistance to water, both in static and dynamic conditions up to temperatures of 200°C. Beyond 250°C beryllium and its alloys are prone to some pitting corrosion. Chlorides present in water increase the corrosion rate quite heavily. Beryllium however shows no sign of stress corrosion.

Beryllium is attacked by dilute HCl, H_2SO_4 and HF at room temperatures but has some resistance to most organic acids. Dilute nitric acid oxidises the surface of beryllium somewhat and for this reason the rate of corrosion is quite low. Concentrated nitric acid violently attacks beryllium and its alloys.

The metal has some resistance to dilute and very dilute caustic solutions but reacts fast with moderately concentrated sodium and potassium hydroxide solutions, particularly at elevated temperatures.

Beryllium has excellent resistance to oxygen even at high temperatures. It is, however, attacked by sulphur, selenium, arsenic and phosphorous when these reagents are hot.

A feature of beryllium is its excellent stability when in contact with molten metals. This is because of the readiness with which the surface is coated with beryllium oxide, BeO, which has good resistance to molten zinc, cadmium, etc.

4.3 Chromium

Chromium has a density of 7·10 kg/dm^3 and a melting-point of 1,830°C. Its main use is as an alloying element for steel, etc., and for surface coatings. It is therefore dealt with in Chapters 6 and 7.

4.4 Cobalt

Cobalt is seldom used on its own but its alloys are becoming quite important structural materials. Its density is 8·9 kg/dm^3 and its melting-point 1,480°C. The main alloys used are the following:

    Cobalt–chromium–nickel
    Cobalt–chromium–nickel–molybdenum
    Cobalt–chromium–nickel–tungsten
    Cobalt–chromium–nickel–tungsten–titanium
    Cobalt–chromium–iron
    Cobalt–chromium–molybdenum
    Cobalt–chromium–tungsten

Cobalt alloys have excellent resistance to water, including water with a high salt concentration, at all temperatures. They are also excellent against all organic acids, as well as sodium and potassium hydroxide solutions at high concentrations and at elevated temperatures.

The resistance of cobalt alloys against inorganic acids is poor and contact with even dilute acids is therefore inadvisable.

These alloys are strongly subject to high-temperature corrosion by oxygen, steam, carbon dioxide and sulphur dioxide. Sulphur dioxide is particularly bad and corrosion rates in excess of 2,500 g/m².day have been reported for cobalt alloys subjected to SO_2 gas at 1,000°C.

Cobalt alloys are widely used for handling molten materials such as molten zinc, sodium and potassium, which have little effect on the alloys although these materials tend to attack many other metals.

4.5 Copper

Copper is noble to hydrogen in the electrochemical series and mainly because of this shows much less tendency to corrode than most other metals. It is very resistant to attack by even the most polluted atmospheres. It forms a continuous and self-healing oxidised skin on its surface which does not absorb moisture and offers strong resistance to further attack. When exposed to the action of air or water for long periods, this skin is converted into complex sulphates and carbonates, popularly known as 'patina'. The green coating often seen on copper roofs is basic copper sulphate derived from traces of SO_2 and SO_3 in the air.

When polished copper is exposed to industrial atmospheres, particularly those containing sulphides, surface tarnishing takes place and the colour changes progressively from bright red to reddish green and finally to dull black.

The ultimate destruction of copper on exposure to even the most polluted atmosphere is extremely slow and does not seem to vary with different grades of copper. No pitting takes place even under the most drastic conditions.

CORROSION IN NATURAL WATERS

Copper is highly resistant to both fresh water and sea-water, the actual corrosion rate being of the order of 0·5–1 g/m².day in quiet sea-water and about double this value in moving, highly aerated sea-water. Copper surfaces always remain free from fouling organisms because the traces of cupric ions released poison such living matter.

The resistance of copper to corrosion is governed by the stability of the protective oxide film on the surface. The rate of corrosion of copper in hard, fairly stagnant waters is very low. Rapidly flowing waters, however, especially if they contain dissolved carbon dioxide or organic acids (humic acids, etc.) tend to dissolve the surface films and thus induce impingement corrosion on the surface of the copper. Unlike the corrosion of iron and steel, neither rust encrustation nor deep pitting are usual. Corroded copper simply looks worn, often with undercut grooving (see Fig. 4.4).

Hot softened water, which contains $NaHCO_3$, is particularly harmful to copper piping. Pitting corrosion can then occur by differential aeration, due to the deposition of dirt or rust at certain points of the copper surface, or when the water contains manganese salts.

Copper is extremely suitable for pipes carrying either hot or cold crude water, or even sea-water, but is unsuitable for the transportation of high-velocity aerated waters. Waters that are rich in oxygen and carbon dioxide, but have low contents of calcium and magnesium ions, should not be pumped along copper pipes at speeds in excess of 1·5 m/sec. Provided the temperature is less than 65°C, and oxygen and carbon dioxide contents are restricted, it is possible to pump moderately hard water along copper pipes at speeds of up to 3 m/sec without corrosion becoming excessive.

Direction of water flow

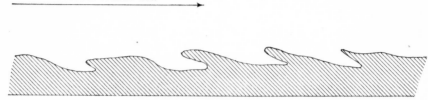

Fig. 4.4. Impingement attack of copper by flowing salt water

CHEMICAL RESISTANCE OF COPPER

Copper resists deaerated hot or cold dilute sulphuric acid, phosphoric acid, acetic acid, hydrochloric acid and many other non-oxidising acids. It also offers good resistance to fatty acids, sulphite solutions and most salts, with the exception of oxidising heavy metal salts such as $FeCl_3$ and $Fe_2(SO_4)_3$.

Copper is attacked by boiling concentrated hydrochloric acid solution even in the absence of oxygen.

It reacts quickly with all oxidising acids. Even very dilute nitric acid attacks copper rapidly. Concentrated sulphuric acid and aerated non-oxidising acids including carbonic acid (in the presence of dissolved oxygen) all corrode copper.

Ammonia forms complex ions with copper:

$$Cu^{2+} + 4NH_3 \rightarrow [Cu\,(NH_3)_4]^{2+}$$

When ammonia reacts with copper the result is intergranular corrosion, causing stress cracking. Great care must be taken that copper is never used in constructions that use ammonia in any form (refrigeration circuits) or even substituted ammonias (amines).

Copper is also readily attacked by acid chromate solutions, mercury salts, perchlorates and persulphates.

CORROSION OF BURIED COPPER PIPES

Experiments were performed regarding the relative life of copper, lead, mild steel and cast iron pipes in various types of soils. The results are shown in Table 4.1. This, however, does not by any means tell the whole story. Copper shows very little pitting—depths are never in excess of about three times the average depth of corrosion. Lead pits badly, and in all cases depths of the deepest pits can be between 15 and 400 times the average depths quoted in the table.

TABLE 4.1   Average surface corrosion rate, g/m^2.day

Type of soil	Copper	Lead	Steel	Cast iron
Cinders	1·0	2·5	5·3	Destroyed
Tidal marsh	0·5	0·02	1·2	1·4
Peat	0·6	0·05	1·5	2·1
Clay	0·35	0·03	Destroyed	Destroyed
Sandy loam	0·15	0·02	0·16	0·08
Sand	0·01	0·015	0·20	0·15
Loam	0·02	0·15	0·45	0·80
Silt loam	0·10	0.25	2·0	3·8

In the case of steel, pitting can amount to about 6–8 times the average depth of corrosion, whereas in the case of cast iron 10–12 times the average depth is often observed. As the failure of a buried underground pipe is not usually due to the thickness of average corrosion, but due to leakage caused by puncturing, the resistance of copper to bad pitting is obviously a strong point in its favour.

Copper corrodes worst in soils with very poor drainage, whereas in well drained soils corrosion is never excessive. Wet soils with high chloride and sulphate contents attack copper most and in such cases copper pipes should be protected by the use of bituminous tapes. It is suggested that copper pipes, coated by hessian and bitumen have an infinite life in even the most corrosive soils. Bacterial action, which is of such importance with regard to corrosion processes of buried iron and steel pipes can be discounted altogether with buried copper pipes because of the strongly bactericidal action of cupric ions in solution. In any but the most corrosive soils, copper piping needs no protection. For this reason copper pipes are being used as standard for many underground services.

CORROSION OF COPPER ALLOYS

The common copper–tin alloys contain between 1·5 and 4% Sn and have virtually the same general corrosion properties as pure copper, although the low tin bronzes have poor stress corrosion resistance. Bronzes with tin contents in excess of 5% are, in general, better than pure copper from the point of view of corrosion resistance.

Brasses, which are copper–zinc alloys, have good resistance to normal impingement attack by various chemicals—better in fact than copper on its own. On the other hand, brasses, especially those with a zinc content above about 25–30%, become prone to the two phenomena of dezincification and stress corrosion cracking.

G

DEZINCIFICATION

Brasses with a zinc content below about 15% are usually quite immune, and rates of dezincification of brasses with less than 30% zinc are quite low. At 40% zinc content the rate of dezincification can become very rapid (see Fig. 4.5). What happens is that traces of copper go into solution as cupric ions and then react with the zinc contained in the brass:

$$Cu^{2+} + Zn \rightarrow Cu + Zn^{2+}$$

The copper is deposited as fine copper dust, which readily dissolves once more in any electrolyte present to form more cupric ions. The zinc is leached out of the brass leaving behind a highly porous mass of copper with negligible mechanical strength.

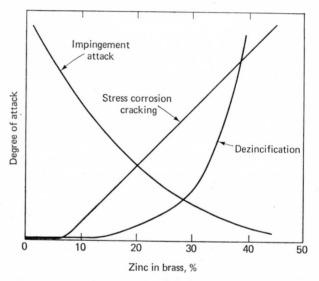

Fig. 4.5. Effect of the zinc content of brasses on (a) impingement attack; (b) dezincification; (c) stress corrosion cracking

Dezincification usually starts when water rich in chlorides and sulphates such as sea-water is in contact with high zinc brass. Brass condenser pipes using sea-water are particularly liable to dezincification. This may take place in layers which are approximately parallel to the surface of the metal, or in the form of plugs with their axes at right angles to the surface (see Figs. 4.6 and 4.7). Often the external appearance of dezincified brass is unchanged, and the first indication of damage is when the brass pipe splits right along its length.

Dezincification is also aided by high temperatures, the presence of stagnant acid solutions and porous inorganic scale formation. By alloying the brasses with about 1% tin and traces of arsenic, antimony and phosphorous, dezincification can be reduced.

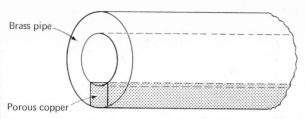

Fig. 4.6. Plug-type dezincification—causes splitting of the pipe along its length

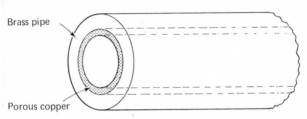

Fig. 4.7. Layer-type dezincification

STRESS CORROSION CRACKING

Brass is prone to stress cracking in the presence of traces of ammonia, amines and oxides of nitrogen, when free oxygen and moisture are also present. Cracks are usually along grain boundaries, although transgranular cracking can occur if stresses are strong. Annealed brass shows little tendency to corrosion cracking, and for this reason the liability of cold-worked brasses to stress cracking can be markedly reduced by heating samples to 300°C for an hour. Hydrogen sulphide acts as an inhibitor as it reacts with free available oxygen. Finally, it is possible to reduce liability to stress corrosion cracking by the use of either an impressed current, or by coating the brass with zinc, which acts as a sacrificial anode.

Modern practice avoids damage of condenser tubes in sea-water by the use of cupro-nickel alloys (20% Ni, 80% Cu) of aluminium brass (22% Zn, 2% Al, 0·04% As, rest Cu). The former is more expensive but is to be preferred to the latter because aluminium brass is prone to pitting attack in stagnant sea-water.

4.6 Gold

Gold is one of the so-called 'noble metals' and has a density of 19·3 kg/dm³ and a melting-point of 1,063°C. It is soft and ductile, and hardly subject to work-hardening at all. When used industrially it is too weak and expensive to be used on its own, and is usually employed in the form of a thin lining or electrodeposit on base metals.

Provided gold does not form a complex, or is used in contact with very strongly oxidising materials, it is totally insoluble at all pH values. The

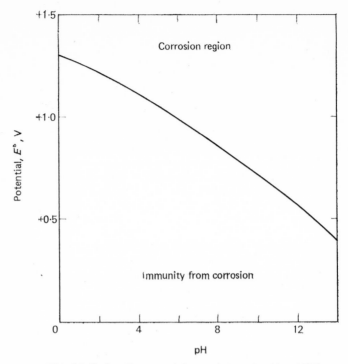

Fig. 4.8. Redox diagram of the corrosion of gold at 25°C

redox potential needed to make gold corrode at different pH values is shown in Fig. 4.8. Passivity of the gold surface is caused only at redox values above about 2·0 V and is therefore of no importance.

Gold is, however, dissolved when complexes are formed, which have lower potential barriers than the common Au ──────→ Au⁺ ionisation. This is particularly the case when gold is brought into contact with alkaline cyanide solutions. Some of the $E°$ values of gold ionisations and various gold complex formations are as follows:

$$
\begin{aligned}
\text{Au} &\to \text{Au}^+ + e & E° &= -1\cdot68 \text{ V} \\
\text{Au}^+ &\to \text{Au}^{2+} + e & E° &= -1\cdot29 \text{ V} \\
\text{Au}^{2+} &\to \text{Au}^{3+} + e & E° &= -1\cdot29 \text{ V} \\
\text{Au} + 2\text{Br}^- &\to (\text{AuBr}_2)^- + e & E° &= -0\cdot96 \text{ V} \\
\text{Au} + 4\text{Cl}^- &\to (\text{AuCl}_4)^- + 3e & E° &= -1\cdot0 \text{ V} \\
\text{Au} + 2\text{CN}^- &\to [\text{Au(CN)}_2]^- + e & E° &= +0\cdot6 \text{ V}
\end{aligned}
$$

It can be seen, therefore, that the main reagents which attack gold are mixes combining complex formation with oxidising properties. The reason aqua regia (a mix of nitric and hydrochloric acids) attacks gold is because of the formation of the (AuCl₄)⁻ complex. Similar reasons underly the ready attack of gold by cyanide solutions, with the formation of [Au(CN)₂]⁻.

4.7 Iridium

Iridium has a density of 22·4 kg/dm³ and a melting-point of 2,442°C. This metal is usually alloyed with osmium and platinum and a common alloy is one containing 40% Ir and 60% Pt. Such an alloy is very hard but still workable, and has excellent anticorrosion properties.

Iridium is by far the most noble of all the noble metals and has far better corrosion-resistance properties than any other metal known. Only ruthenium is at all comparable. Iridium has complete resistance to all acids and alkalis, at all concentrations and temperatures, including even such very drastic reagents as aqua regia and hydrofluoric acid. It is totally unattacked by all halogens, salt solutions and acid gases.

In air iridium is slightly oxidised at temperatures above 600°C, but the oxide formed is reconverted to iridium metal at about 1,000°C.

Only molten cyanides attack iridium, and even these reagents do so to a lesser extent than they do other noble metals.

At 700°C iridium is attacked at the rate of 180 g/m².day by potassium cyanide and to the extent of 450 g/m².day by sodium cyanide. The reason for this attack is the formation of iridium–cyanide complexes.

4.8 Lead

Lead has a density of 11·344 kg/dm³ and a melting-point of 327°C. It is widely used in industry as a corrosion-resistant material, although it is not particularly good for such purposes at elevated temperatures. One of the most useful properties of lead is its ability to absorb nuclear and X-ray radiation, and it is therefore widely used in such fields.

The most common lead alloys contain 4–12% antimony. Such alloys are unsuitable for use above about 120°C, as they soften appreciably above this temperature.

Lead accumulators are filled with sulphuric acid solutions which vary in concentration between 37% w/v when fully charged to 20% w/v when fully discharged. Antimony alloys of lead have greater resistance to sulphuric acid than lead alone and such alloys are therefore used for making the plates used in lead accumulators.

Lead is also applied by hot dipping to steel, to improve its corrosion resistance (Terne plating).

RESISTANCE OF LEAD TO ACID

The single main criterion as to whether lead is resistant to a given acid is the solubility in water of the appropriate lead salt produced. Table 4.2 gives the solubility of a number of lead salts in water at 25°C, given as parts per million of the lead salt in water, under equilibrium conditions.

It becomes obvious that lead must never be used in conjunction with nitric acid, acetic acid and formic acid, as these dissolve it very rapidly indeed. The resistance of lead to hydrochloric acid is also not too good, but the solubility of the other lead salts is sufficiently slight for lead to show good resistance to the appropriate acids, particularly at low temperatures.

Lead is widely used for handling sulphuric acid. When the concentration of the acid is below 5%, the corrosion rate is fairly rapid because of the ready solubility of the lead sulphate in water (see Fig. 4.9). At higher concentrations, the solubility of lead sulphate is restricted, and therefore the resistance is better. Lead can be used with 80% sulphuric acid at temperatures up to 100°C and at room temperature for sulphuric acid with a concentration of up to 94%. If higher concentrations or higher temperatures are used, lead is rapidly destroyed. Slight traces of impurities in either the acid or in the lead increase the rate of corrosion of lead in sulphuric acid. Agitation of the solution also has considerable effects on the rate of corrosion as can be seen in Fig. 4.10. Lead has good resistance to H_2SO_3.

TABLE 4.2

Lead salt	Solubility, ppm, at 25°C in water
Nitrate	540,000
Acetate	450,000
Formate	16,000
Chloride	6,700
Fluoride	650
Hydroxide	155
Sulphide	120
Sulphate	43
Oxalate	1·6
Oxide	1·5
Carbonate	1·0
Phosphate	0·15
Chromate	0·06

Lead withstands hydrochloric acid in a hydrogen atmosphere, but the rate of corrosion when oxygen is present is roughly 10 times as fast. For example at 25°C the rate of corrosion (grammes per square metre per day) is as shown in Table 4.3.

TABLE 4.3

	Oxygen atmosphere	Hydrogen atmosphere
4% HCl solution	130	13·5
20% HCl solution	880	88·0

The same is also found with regard to the solubility of lead in other acids. As is to be expected, the resistance of lead to acetic acid and other organic acids is not good. There exists a very considerable difference in the rate of corrosion of lead in contact with organic acids, depending upon whether the lead is quietly immersed, aerated or sprayed with the acid.

At 25°C, 0·75N acetic acid has the following effect on lead:

Quiet immersion	:	corrosion rate =	9 g/m^2.day
Aeration	:	corrosion rate =	36 g/m^2.day
Spray	:	corrosion rate =	216 g/m^2.day

As is to be expected, lead has no resistance at all to nitric acid even at very low concentrations and temperatures. In fact, lead used for radiation shielding or X-ray shielding is often corroded. The reason given is that under the action of short-wave rays, oxides of nitrogen are formed and these react with moisture to form nitric acid, which in its turn attacks lead.

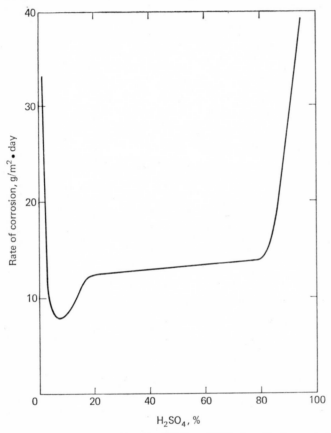

Fig. 4.9. Rate of corrosion of lead by H$_2$SO$_4$ at 25°C when the flow-rate is 1 m/sec

RESISTANCE TO OTHER AGENTS

Lead is readily attacked by soft, distilled or deionised water in the presence of air, as the lead then goes into solution as soluble Pb(OH)$_2$. The limit of lead content in domestic water is specified as 0·1 ppm, as lead poisoning can occur if more lead than this is present in drinking water. One must therefore *never* pipe soft water through lead piping as this limit will then

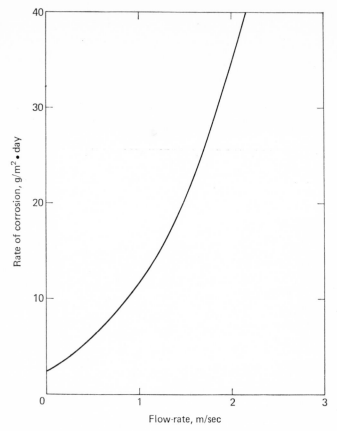

Fig. 4.10. Rate of corrosion of lead in 20% H₂SO₄ at 25°C at varying flow velocities

certainly be exceeded. In areas where water is naturally soft, some calcium bicarbonate is usually added to town's water to ensure that the lead pipes of the supply system are lined with a protective film. The presence of nitrates, nitrites and ammonia in the water all increase the rate of corrosion of lead, as does the presence of free carbon dioxide and oxygen in the water carried. Pure, gas-free, deionised or distilled water does not attack lead nor has chlorination up to 0·002 ppm any appreciable effect upon the plumbosolvent power of water.

Although lead has poor resistance to soft or distilled water it withstands hard water and especially sea-water extremely well. The corrosion rate by sea-water has been found to be less than 0·1 g/m².day, or about 0·006 mm/year. Lead is also extremely resistant to salt solutions, which have virtually no effect upon it at temperatures below 80°C. Solutions of oxidising agents such as potassium dichromate and potassium permanganate can be kept in lead containers for unlimited periods of time.

Lead resists dry chlorine up to about 100°C and wet chlorine up to about 80°C. Its resistance to bromine vapour or to hydrogen fluoride is poor, particularly at elevated temperatures.

Organic chemicals generally have little effect upon lead, with the exception of some of the acids and also benzaldehyde and nitrophenols, which must never be kept in lead containers or passed through lead piping. Lead is also resistant to weak alkalis such as ammonia, sodium carbonate etc. Strong alkalis like sodium hydroxide and potassium hydroxide attack lead, particularly if the solution is air agitated. The reason for this is that strong caustic solutions form soluble plumbates with lead oxide:

$$2NaOH + PbO \rightarrow Na_2PbO_2 + H_2O$$

Lead has superb resistance to even the most polluted atmospheres. Tests carried out on various metals to compare their resistance to an industrial atmosphere showed that the relative weight losses after seven years' exposure were as follows:

Mild steel	2,609
Wrought iron	1,703
Zinc	137
Nickel	120
Nergandin brass	100
Copper	47
Aluminium	29
Tin	18
99·96% lead	22
98·4% lead, 1·6% antimony	3

Lead also has extremely good high-temperature corrosion resistance to air and other gases.

SOIL AND GALVANIC CORROSION OF LEAD

Buried lead objects corrode at a rate markedly dependent upon the nature of the soil with which they are in contact. Humus, cinders and heavy clays are by far the worst soils from the point of view of the corrosion of lead. Sandy soils, in general, have little effect. The worst action is the very heavy pitting that occurs, and it is not unknown for pits to exceed 400 times the average depth of corrosion penetration. In heavy clays the average corrosion may amount to 7 μm/year with pit depths of 2·5 mm. Lead piping, cable covers etc. can become subject to stray current corrosion. Protection of such lines is then usually carried out by means of electrical drainage (see Chapter 3). Lead also tends to corrode when in contact with rotting timber due to the fact that organic acids are liberated, which attack the metal.

Copper in contact with lead makes the latter anodic and therefore increases its rate of corrosion. However, in most media this simply means that the protective film formed on the lead is thicker and therefore the rate at which lead goes into solution soon falls to a very low level, provided that the lead salt has a low solubility in the medium in question. Aluminium and lead should never be used together since the former is very rapidly attacked under such circumstances. Lead is often alloyed with tellurium, antimony and some other metals to increase the strength of the material. In general the properties of such alloys from the corrosion point of view differ little from those of chemically pure lead.

4.9 Magnesium

Magnesium is a metal that is even lighter than aluminium, having a density of only 1·70 kg/dm³ against 2·71 for aluminium. Its melting-point is 651°C and its electrical potential for $Mg \rightarrow Mg^{2+} + 2e$ is

$$E° = +2·34 \text{ V at } 25°C$$

Magnesium is usually employed in alloyed forms. Both casting and wrought alloys are made, and the most common alloying metals are zinc and aluminium. Alloying percentages of up to 5·5% Zn and up to 9·5% Al are used. In addition, small quantities of zirconium, thorium, beryllium, manganese and mixtures of rare earths are added.

The casting alloys MSR A and B contain 2·5% silver, 1·7–2·5% rare earths and 0·6% zirconium.

ATMOSPHERIC CORROSION

Magnesium is tolerably resistant to ordinary atmospheric exposure but is prone to stress cracking in damp air, which also causes pitting corrosion. In industrial atmospheres magnesium corrodes at a rate of 0·4 g/m².day and in marine atmospheres, where surfaces are likely to be wetted by salt water, corrosion rates for magnesium alloys can be as high as 2 g/m².day. High-purity metal does not corrode as much as this. Magnesium, like aluminium, is particularly subject to crevice corrosion and should not be used externally except in the heavily chromated and painted state. The only magnesium alloy that can normally be used for external purposes without additional surface protection is one containing 3% Zn, 1·5% Mn and 0·7% Zr.

Magnesium is strongly affected by galvanic corrosion because it is likely to be anodic to nearly every other metal with which it is in contact but it can be coupled safely with a 95% aluminium 5% magnesium alloy (see Fig. 4.11). When coupling magnesium with other metals the adverse effects can be reduced by alloying it with zinc. In solution chromates, vanadates and sulphides serve as corrosion inhibitors. Tiny quantities of impurities in magnesium often cause heavy localised corrosion, the reasons again being galvanic phenomena.

The tolerance limit for iron impurities in magnesium is as low as 0·02% and for nickel 0·0005%. The permissible limit of copper in magnesium is 0·1%.

RESISTANCE TO ACIDS AND ALKALIS

Hydrofluoric acid does not attack magnesium except at high temperatures, due to the formation of an insoluble fluoride on the surface of the metal.

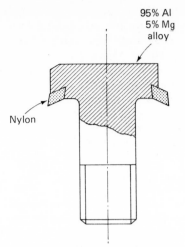

Fig. 4.11. 'Galvanut' used for jointing magnesium sections *(By courtesy of Magnesium Electron Limited)*

Except for hydrofluoric acid, magnesium has a very poor resistance to nearly all inorganic acids, both in the dilute and the concentrated states.

Boiling 20% chromic acid removes corrosion products from the surface of magnesium metal without attacking the metal itself. However, it is absolutely essential that traces of chloride ions be absent, otherwise corrosion is very rapid. This is ensured by adding a tiny quantity of Ag_2CrO_4 to the solution; this reacts with chloride ions to form insoluble silver chloride.

Magnesium has quite good resistance to cold and dilute alkali solutions, because a coherent and highly protecting film is produced on the surface. Magnesium is far better in this respect than aluminium, which dissolves rapidly in alkaline solutions.

CORROSION IN WATER AND SALT SOLUTIONS

Magnesium is not attacked appreciably by fresh water in the cold. The corrosion rate at 20°C is only 0·8 g/m^2.day due to the formation of a protective $Mg(OH)_2$ film on the surface. As the temperature of the water in contact with the metal is increased, the rate of attack increases. The corrosion of magnesium and magnesium alloys when in contact with water at its boiling-point is between 1·5 and 6·5 g/m^2.day depending upon the nature of the alloy.

Neutral or acidic solutions of salts attack magnesium and its alloys badly. The only salts with which magnesium can be kept safely in contact are sodium and potassium chromates, fluorides, phosphates and nitrates. The reaction products of all these substances with magnesium are insoluble in water. The chlorides, bromides, iodides and sulphates of all metals have a highly corrosive action on magnesium. The salts of metals such as aluminium, iron and copper are much worse than sodium and potassium salts, because of the liability of cathodes being formed on the surface of the magnesium. The reason for the corrosive effect of halides and sulphates is that magnesium salts of these acid radicles are soluble in water. The rate of corrosion in all cases is higher when the solution is acid as well, i.e. when the pH is low. But even if the solution is slightly alkaline, magnesium, unlike iron, can release hydrogen on the cathodic side of the corrosion cell. This can cause blistering when magnesium is coated over by a paint film which is penetrated by some of the salt solution. Magnesium is attacked badly by sea-water which produces heavy pitting. Exposure of magnesium to salt solutions is also responsible for very bad intergranular stress corrosion.

RESISTANCE TO ORGANIC COMPOUNDS

Magnesium has good resistance to many organic materials. It should, however, not be used in conjunction with low-molecular-weight organic acids, although it has good resistance to higher fatty acids. Anhydrous methanol attacks magnesium and organic halides form the well known Grignard reagents with magnesium metal. For example, magnesium together with ethyl chloride forms the following compound:

$$Mg + C_2H_5Cl \rightarrow C_2H_5\ MgCl$$

SURFACE PROTECTION OF MAGNESIUM

Chemical and electrochemical treatments
The metal can be oxidised to form a stable and coherent oxide film on the surface. This is usually done by immersing the metal in a chromate bath or by an anodising treatment. The magnesium object is made the anode in a bath which contains a solution of ammonium fluoride with free hydrofluoric acid.

Paint coatings
Magnesium is commonly protected by various coherent paint coatings. The best method is undoubtedly the use of a high-temperature epoxy film. Magnesium, when protected in this way, can be used even in contact with salt water.

Literature sources and suggested further reading

1 LA QUE, F. L., and COPSON, H. R., *Corrosion resistance of metals and alloys*, Reinhold, New York (1963)
2 *Copper underground. Its resistance to soil corrosion*, Copper Development Association, London (1967)
3 *Copper data*, Copper Development Association, London (1966)
4 *An introduction to aluminium and its alloys*, Aluminium Federation, London (1968)
5 *Anodic oxidation of aluminium and its alloys*, Aluminium Federation, London (1968)

6 GREATHOUSE, G. A., and WESSEL, C. J., *Deterioraton of materials*, Reinhold, New York (1954)

7 BOOKER, C. J. L., *Corrosion of metals and alloys*, Translated from the Russian, National Lending Library for Science and Technology (1967)

8 MANN, J. Y., *Fatigue of materials*, Melbourne University Press, Melbourne (1967)

9 'Corrosion chart', *Chemical Processing* (December 1969)

10 *Protection of magnesium rich alloys against corrosion*, Ministry of Technology, London, DTD 911C (January 1967)

11 MAGNESIUM ELECTRON LIMITED, Technical brochures

12 LEAD DEVELOPMENT ASSOCIATION, Technical brochures and information

Chapter 5   Corrosion of Non-ferrous Metals—II:

Molybdenum, Nickel, Niobium, Palladium, Platinum, Silver, Tantalum, Tin, Titanium, Tungsten, Uranium, Vanadium, Zinc, Zirconium

5.1 Molybdenum

Molybdenum is nowadays mainly used for the construction of high-temperature furnaces for use with protective atmospheres and vacuum. It also has certain uses in the chemical industry where it is used for corrosion-resistant components. Its density is $10 \cdot 2$ kg/dm³ and its melting-point 2,620°C.

Molybdenum is a steely grey metal which has good resistance to hot and cold concentrated hydrochloric, hydrofluoric and sulphuric acids. Nitric acid below 100°C, both dilute and concentrated, has little effect, but at temperatures above 200°C molybdenum is attacked very rapidly by both concentrated nitric and sulphuric acids.

The metal has good resistance to cold and hot aqueous alkalis but is attacked by fused caustic soda and caustic potash. Chlorine, bromine and iodine have no effect at all normal temperatures but do attack molybdenum at red heat. Fluorine on the other hand, tends to attack molybdenum even at room temperature. There are considerable dangers when molybdenum is used in conjunction with lead dioxide (PbO_2) and molten oxidising salts, such as KNO_3, KNO_2, Na_2O_2, $NaNO_3$ or $NaNO_2$, particularly if the molybdenum is in powder form; the reactions that then take place can be quite violent.

Molybdenum is quite stable in oxygen or air at normal temperatures, but surface oxidation, which starts at 400°C, becomes rapid above 500°C. Water, salt solutions and organic materials have no effect on molybdenum whatever, nor is molybdenum attacked by such gases as sulphur dioxide, hydrogen sulphide, carbon monoxide, carbon dioxide, or ammonia gas, below 1,000°C. Ammonium hydroxide attacks molybdenum slowly.

Molybdenum is, however, attacked by many reagents at red heat. Even steam corrodes molybdenum at above 650°C.

5.2 Nickel

Nickel has a density of $8 \cdot 88$ kg/dm³ and a melting-point of 1,450°C. Electrochemically it comes between iron and copper, with a value of

$$\text{Ni} \rightarrow \text{Ni}^{2+} + 2e \qquad E° = +0·25 \text{ V}$$

Because of the high degree of polarisation of hydrogen on nickel surfaces the metal is not readily dissolved in mineral acids in the absence of a depolariser. This is one of the reasons why there is such a very considerable difference in the rate of corrosion in oxygenated and non-oxygenated solutions (see Fig. 5.1). Under oxidising conditions nickel forms a coherent and protective oxide film on its surface.

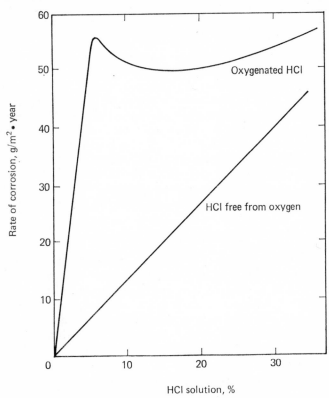

Fig. 5.1. Corrosion of nickel by oxygenated and deoxygenated HCl

The main forms of nickel used in commerce are cast nickel (1·6% Si, 0·8% C and 97·6% Ni) and high carbon grade nickel (1·5% C and 98·5% Ni). Many other nickel alloys are also being used, the main alloying elements being Cu, Fe, Mn, Si and C. In general, the corrosion-resistance properties of most of these alloys are the same as those of pure nickel. The metal is virtually immune to stress corrosion.

RESISTANCE AGAINST AQUEOUS SOLUTIONS

Nickel has excellent corrosion resistance against all forms of water including sea-water. In the absence of oxidising agents or free oxygen, nickel has superb resistance to all organic acids in aqueous solution. As nickel salts are non-toxic, the metal is very widely used for equipment intended for handling food and drugs.

Nickel has good resistance to all salt solutions except oxidising salts, particularly alkaline ones such as sodium hypochlorite, NaOCl. Only very dilute solutions of oxidising salts can be safely handled in nickel equipment. It is attacked slightly by hot steam condensate if this is impregnated with carbon dioxide.

RESISTANCE AGAINST ACIDS AND ALKALIS

Nickel has good resistance against organic acids, HCl, H_2SO_4, and H_3PO_4 and other non-oxidising acids including even hydrofluoric acid, provided these are not too concentrated, too hot and, above all, do not carry too much dissolved oxygen. The latter is the most important criterion in all cases due to the ready polarisation of nickel surfaces. The metal has no resistance whatever to nitric acid except when this acid is cold and very dilute. Its resistance to alkalis is excellent even at elevated temperatures and high concentrations. Nickel can withstand molten sodium hydroxide but only if a low-carbon alloy is used. It has good resistance to low concentration ammonia solutions and anhydrous ammonia. Concentrated ammonia-water mixes in the presence of free oxygen cause corrosion.

OTHER CORROSION PROPERTIES

Nickel has excellent resistance to atmospheric corrosion but tends to fog slightly, which detracts from its visual appearance. For this reason nickel is often covered by a minutely thin film of chromium. Nickel has excellent resistance against halogens, including fluorine gas, when these gases are dry but less in the presence of moisture. Nickel also withstands such usually corrosive gases as ClF_3, HF, NOCl, etc., again, only when these gases are dry. It also has good resistance to nearly all organic compounds.

GALVANIC COUPLES WITH NICKEL

The worst possible combination of this kind is the nickel–aluminium couple because aluminium is very rapidly destroyed. Nickel–iron couples should also be avoided where possible. It has been found that corrosion attack on iron is often increased ten-fold if the iron is connected electrically to nickel. Nickel–tin, nickel–zinc, and nickel–lead couples are usually satisfactory in neutral solutions and nickel can also be coupled without any ill-effect to copper, brass or bronze.

5.3 Niobium

This metal, usually referred to in Great Britain as 'niobium', is known in the United States as 'columbium'. Because it has a low neutron-absorption cross-section and is able to resist molten sodium and sodium–potassium alloys, it is finding increasing use in the field of nuclear energy, particularly in the design of fast reactors. Niobium has a density of 8·55 kg/dm³ and a melting-point of 2,470°C. It has excellent resistance to water and all aqueous solutions of salts up to quite elevated temperatures.

At room temperature niobium resists dilute and concentrated acids, including aqua regia. At a temperature of 60°C aqua regia only dissolves niobium at a rate of about 0·8 g/m².day. The resistance of niobium to oxidising acids is particularly good as the surface is coated with a coherent film of Nb_2O_5. The corrosion resistance of niobium to acids at higher temperatures is still quite good but is markedly improved by alloying the metal with molybdenum, titanium, vanadium, zirconium and molybdenum to form either double, triple, or quadruple alloys. These alloys usually have extremely coherent oxides on their surfaces which are able to withstand hot and concentrated mineral acids.

Alkalis tend to corrode niobium and its alloys, because they form water soluble niobates. A 5% KOH solution dissolves niobium at the rate of about 4·5 g/m².day at 20°C, whereas at temperatures approaching the boiling-point of the solution the rate of attack is very high indeed. Sodium hydroxide solutions are even more active. Caustic solutions also tend to embrittle niobium.

HIGH-TEMPERATURE CORROSION OF NIOBIUM

Niobium tends to oxidise at high temperatures and is also readily attacked by other gases such as nitrogen and hydrogen (see Fig. 5.2). Steam attacks niobium at high temperatures too, and at a temperature of 400°C, high-purity niobium loses 1 g/m².day. Commercial niobium in these conditions has a corrosion rate of 3·5 g/m².day and further attack is then very much accelerated. Niobium can be used in conjunction with many molten metals.

Table 5.1 gives the maximum temperatures at which it is possible to use niobium with a given metal.

TABLE 5.1

Metal	Temperature, °C
Lead	1,000
Sodium	900
Sodium/potassium alloy	900
Lithium	900
Gallium	450
Mercury	300
Bismuth	300
Lead–bismuth alloy	Poor resistance
Zinc	Poor resistance

H

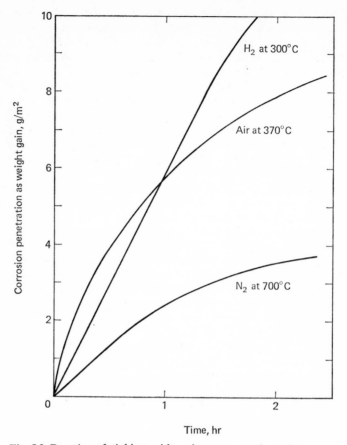

Fig. 5.2. Reaction of niobium with various gases at elevated temperatures

5.4 Palladium

Palladium has a density of 12·0 kg/dm³ and a melting-point of 1,552°C. Although it is generally referred to as a 'noble metal', palladium does not have unduly good corrosion resistance properties.

It is unattacked by water or moist air and has good resistance to a number of non-oxidising acids at room temperature, but is readily attacked by hot and concentrated non-oxidising acids as well as nitric acid at any but the lowest concentrations.

For example, 98% sulphuric acid at 300°C attacks palladium at a rate of 1,000 g/m².day. Palladium corrodes rapidly in the presence of moist chlorine, bromide or iodine and has only fair resistance to completely dry halogens.

Palladium has some resistance to salt solutions with the exception of ferric chloride, cyanides and hypochlorites, and withstands aqueous solutions of alkalis fairly well at all temperatures. It is stable to fused alkalis up to about 550°C but beyond this temperature palladium is attacked rapidly.

Of all the noble metals it has the poorest resistance to fused potassium cyanide which attacks it at 700°C at the rate of about 3,000 g/m².day. Palladium is normally alloyed with platinum and fortunately such alloys tend to resemble platinum rather than palladium with regard to their corrosion-resistance properties.

In air palladium is covered with an oxide film at about 350°C, which is broken down again to metallic palladium once the temperature rises above 800°C.

5.5 Platinum

Platinum is a soft and ductile metal with a melting-point of 1,769°C and a density of 21·4 kg/dm³. It is one of the most corrosion resistant metals known, because its $E°$ value is equal to $-1·20$ V, which makes platinum cathodic to virtually every other metal with the exception of gold. If platinum is made anodic to some other metal—by virtue of an impressed external electric current or due to very strong oxidising agent—it does not readily corrode because its surface then becomes passive (see Fig. 5.3). In fact, platinum can corrode only under the combined conditions of an impressed current or oxidising action to make it anodic within the limits of about 1–1·3 V *and* a pH which is less than zero. It can be appreciated that conditions of this type do not occur very often. Platinum has therefore complete resistance to alkaline, neutral or acid aqueous solutions at virtually all concentrations and temperatures. The only conditions under which platinum can be attacked are, as mentioned before, very acid conditions with a combination of extremely strong oxidising action.

Platinum is therefore attacked by aqua regia (HCl/HNO_3) in the cold and in the hot, and by very hot selenic acid.

Cyanides tend to form complex salts with platinum and for this reason platinum is dissolved by very hot alkali cyanides. The metal is also attacked slightly by chlorine gas and somewhat more vigorously by bromine.

Platinum resists sulphuric acid, nitric acid and hydrochloric at all concentrations and temperatures. It is unattacked by fused sodium and potassium hydroxides up to a temperature of around 600°C. It can be kept in contact with most fused salts too at temperatures which vary between 500°C and 1,100°C, without any appreciable attack occurring. Only ferric chloride corrodes platinum at more moderate temperatures.

THE USE OF PLATINUM AS AN ANODE

Platinum is widely used as an impressed anode to protect structures of other metals. Unlike magnesium and zinc, which are themselves eaten up and act therefore as so-called 'sacrificial' anodes, platinum is protected by a passive film and is hardly attacked at all, even under very heavy

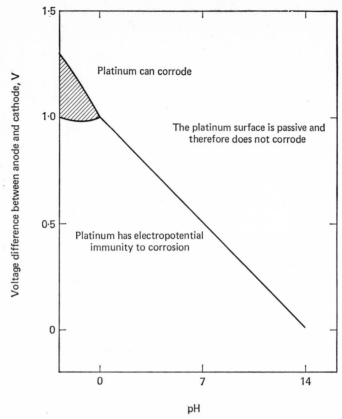

Fig. 5.3. Condition under which platinum can corrode

current densities. The loss of platinum per ampere-year is only 0·0006 g. In actual fact, platinum in concentrated acid solutions such as hydrochloric acid becomes more vulnerable when a.c. is passed than under the effect of d.c.

To protect ships' hulls against corrosion, sheets of a 50% platinum and 50% palladium alloy, 0·25 mm thick and mounted on plastic sheeting are made, which are then attached to the hull to act as anodes. Platinum is also used as a surface layer on sheet titanium for the protection of fixed installations such as steel jetties, when it also acts as a protective insoluble anode.

5.6 Silver

Silver has a melting-point of 1,760°C and a density of 10·5 kg/dm^3. It is a highly cathodic noble metal with

$$Ag \rightarrow Ag^+ + e \qquad E° = -0·79 \text{ V}$$

Industrially it is used in two gradings: 99·95% silver and 99·99% silver. The former is harder, while the latter has better resistance to chemicals.

RESISTANCE TO ACIDS, BASES AND SALTS

Silver behaves well against non-oxidising acids. Its resistance to all organic acids is excellent and it also withstands hydrochloric acid, which produces a coherent chloride film on its surface. This resistance is much reduced when traces of zinc, aluminium or similar metals are present in the silver, due to the fact that the coherency of the silver chloride film is then reduced. Silver is not attacked by sulphuric acid at room temperature, but its resistance to hot concentrated sulphuric acid, particularly if traces of nitric acid or other oxidising agents are present, is not too good.

Nitric acid attacks silver rapidly, even if cold and dilute, and it is very readily dissolved by aqua regia. Although silver fluoride is readily soluble in aqueous solution, silver can be used as a container for hydrofluoric acid at room temperature, at concentrations of up to 40% HF. The presence of dissolved oxygen in this acid, as in any other acid, increases the rate of attack manifold.

Silver is the perfect metal to use in contact with alkalis. It is totally unattacked by sodium hydroxide and potassium hydroxide solutions at any concentration and temperature, or indeed in the fused state. It is widely used for equipment handling concentrated and fused alkalis.

Silver is totally unattacked by water and nearly all aqueous solutions of salts, with the exception of chlorides of heavy metals, such as ferric chloride, particularly in the presence of oxygen or oxidising agents.

RESISTANCE TO OTHER SUBSTANCES

Silver has good resistance at room temperature to chlorine, which forms a protective and coherent silver chloride film on its surface, but is, however, attacked by both bromine and iodine vapour. It tarnishes rapidly in the presence of hydrogen sulphide and other sulphur containing gases (even in air) or in contact with materials containing organic sulphides. The silver sulphide film formed is only very thin and has a good protective action.

5.7 Tantalum

This metal is widely used in the chemical industry in spite of its very high price, because of its excellent corrosion-resistant qualities. Tantalum is a bluish-grey metal which looks somewhat like lead. In fact, it has an even higher density than lead, 16·6 kg/dm^3, and a very high melting-point, 2,996°C.

Because tantalum tends to form surface reaction compounds when heated to red heat in nearly all gases with the exception of the inert gases, it is necessary to cold-work the metal. Luckily, tantalum is very ductile and can easily be worked to form very strong products.

CORROSION RESISTANCE TO ACIDS AND BASES

Tantalum has excellent corrosion resistance to hydrochloric acid, hydrobromic acid and hydriodic acid at all concentrations and temperatures. Its resistance to dilute and concentrated sulphuric acid is also excellent even at very high temperatures.

The metal is attacked badly by hydrofluoric acid, fluosilicic acid and also by hot fuming sulphuric acid in the presence of fluorine.

Care must be taken when tantalum is kept in long-term contact with non-oxidising acids, to prevent hydrogen embrittlement. This can normally be prevented by the presence of small surface areas of platinum, in conjunction with tantalum.

Tantalum shows excellent corrosion resistance to phosphoric acid at all strengths, and to all inorganic salts as well as dilute sodium and potassium hydroxide solutions, right up to their boiling points. It is attacked badly by 40% NaOH as well as by fused alkalis.

HIGH-TEMPERATURE CORROSION

Steam is without effect on tantalum, up to a pressure of 12 bar. Steam at very high pressure tends to attack tantalum slightly. Between 500 and 1,000°C tantalum is vigorously attacked by oxygen, nitrogen, hydrogen, carbon monoxide and carbon dioxide. Films, which do not seem to protect the rest of the metal too well, are produced on the surface of the metal and consequently the products of the reaction tend to penetrate quite deeply. Chlorine and bromine have little effect, except at very high temperatures, but the resistance of tantalum to fluorine gas, even at fairly low temperatures, is not too good.

Tantalum has excellent resistance to many molten metals. The following metals can be kept in the molten state in tantalum vessels at a temperature of up to 1,000°C without appreciable destruction of tantalum taking place: bismuth, lead, lithium, sodium, potassium, mercury, lead–bismuth alloys.

It is inadvisable to keep either molten tin or molten zinc in tantalum vessels, as these metals tend to attack it.

5.8 Tin

Tin is used industrially in the form of the 99% pure metal, as well as in the form of a number of different alloys. The most common of these are:

1 TIN–COPPER–ANTIMONY ALLOYS: 98% Sn, 1% Cu, 1% Sb and 90% Sn, 2% Cu, 8% Sb.

2 SOLDERS: a large number of different solders are used, which employ tin as their main ingredient. They also contain lead as well as smaller quantities of antimony, bismuth, zinc etc.

3 BEARING METALS: a common alloy is one containing 75% Sn, 10% Pb, 10% Sb and 5% Cu.

Tin is also used widely as a coating metal for steel and other base metals.

CORROSION IN ACIDS AND BASES

Tin is badly attacked by nearly all inorganic acids, even at low concentrations, but the rate of corrosion is very markedly dependent upon the ready availability of oxygen. In hydrogen atmospheres the rate of attack of tin by dilute hydrochloric acid and dilute sulphuric acid is fairly low (see Table 5.2).

TABLE 5.2   Rate of corrosion at 25°C, g/m².day

Acid	Oxygen atmosphere	Hydrogen atmosphere
6% HCl	1,350	7
6% H_2SO_4	530	4

As nitric acid is an oxidising acid, it makes no difference whether an oxygen or a hydrogen atmosphere is present. With 6% nitric acid the corrosion rate is, in both cases, 75 g/m².day.

Tin has good resistance to most organic acids, with the exception of hot acetic acid. It withstands ammonia below 20°C, but is readily attacked above this temperature. Its corrosion resistance to other alkaline solutions depends upon whether the pH value of the solution is high enough to dissolve the oxide film which has formed on the surface of the tin. This critical pH value is normally about 10. If the pH value of the alkaline solution in contact with the tin is below this, the tin has excellent resistance and is hardly attacked at all. When the pH is in excess of 10, soluble stannates are formed and consequently the corrosion rate becomes very rapid indeed. The corrosive attack of alkalis on tin can be reduced by two completely different techniques:

1 By the complete removal of oxygen from the solution by the use of Na_2SO_3, hydrazine, etc.
2 By adding a strong oxidising agent such as $KMnO_4$ or $K_2Cr_2O_7$ to the solution; this acts as an inhibitor.

CORROSION IN AIR AND IN AQUEOUS SOLUTIONS

Tin is attacked very slowly by clean and dry air, but even if the atmosphere is humid and polluted, the rate of corrosion still does not exceed 2·5 g/m².year.

Distilled water has no adverse effect at all, but some local corrosion can take place in contact with salt solutions that do not form insoluble stannous salts. These are the chlorides, bromides, nitrates and sulphates, which may cause some pitting. Phosphates, carbonates and bicarbonates have no corrosive effect on tin whatsoever. Tin is also reasonably resistant to sea-water, the rate of corrosion being of the order of 0·05 g/m².day.

When tin is joined to either copper or nickel it becomes anodic and therefore the rate of corrosion is increased. On the other hand, when tin is coupled to aluminium or zinc, it acts as a cathode and corrosion takes place on the other metal. Tin must never be used in conjunction with steel in corrosive environments because under such circumstances the steel becomes liable to very rapid attack.

In general, small quantities of bismuth and antimony added to tin improve its corrosion resistance, whereas the addition of copper and aluminium increases the likelihood of corrosive attack.

SOLDERS

There is a considerable danger of corrosive attack affecting soldered joints because the surface areas of these joints are relatively small, and therefore the anodic reaction penetrates deeply. This particularly affects soldered copper pipes exposed to sea-water. In order to increase the surface area of the anodic part it is advisable to tin the copper along a fair part of its length. There are particular dangers to soldered joints between copper sections in hot water systems. Soldered brass pipes generally do not give much trouble, because the potentials of brass and solder are virtually the same. Solders are cathodic to steel, zinc and cadmium and because under such circumstances the cathodic solder area is small, while the anode has a large surface area, corrosion is no danger. Heavy corrosion is, however, to be expected when aluminium is jointed by lead–tin solder. It is better to use lead–zinc solder instead, if the aluminium is to be used in corrosive places.

As a general rule, the lead content of a solder should be restricted as far as possible in corrosive environments.

Also many troubles found with the corrosion of soldered joints are due to the presence of remnants of flux; these should be removed carefully before the soldered joint is exposed to corrosive conditions. Troubles have been encountered with soldered car radiators during the winter, when ethylene glycol antifreeze was added. To protect the soldered joints, sodium benzoate or borax is always added to antifreeze solutions as an inhibitor.

Finally, it must never be forgotten that bearing alloys are strongly cathodic to steel, particularly in salt water. A tin-bearing alloy must be avoided in conjunction with steel shafts, if the joint is likely to be exposed to sea-water or similar highly corrosive fluids.

5.9 Titanium

Pure titanium is a metal with a density of 4·51 kg/dm³, a melting-point of 1,800°C and somewhat better mechanical properties than mild steel. Titanium is intrinsically rather reactive and immediately acquires a thin coating of oxide as soon as it is exposed to the atmosphere. This coating is the rutile form of titanium dioxide and is responsible for the excellent corrosion resistance of the metal. Provided oxygen is present, the film is self-healing and reforms as soon as it is damaged mechanically.

Titanium is particularly resistant to chemicals that have normally strong oxidising actions. It is totally unaffected by wet chlorine gas and by solutions such as sodium chlorate, chlorite and hypochlorite. It has excellent resistance to sea-water even under high velocity conditions.

Titanium is normally immune to stress corrosion, provided the metal is fairly pure. The only occasions when stress corrosion may occur are when it comes into contact with completely anhydrous nitric acid or methanol. Crevice corrosion with titanium is much rarer than with most metals.

CORROSION IN SEA-WATER

Titanium has been found to be virtually immune to sea-water even under the most unfavourable conditions and at highly elevated temperatures. It is ideal for use in desalination plants where titanium has an infinite life at temperatures up to 125°C, being also completely unaffected by corrosion fatigue. When titanium is used in conjunction with other metals in sea-water it becomes the cathodic member and thus tends to induce more rapid corrosion in the other metal. This particularly applies when titanium is used in conjunction with mild steel, aluminium and various copper alloys, when penetration rates of between 0·2 and 0·7 mm/year can be expected. Titanium can, however, be used in conjunction with aluminium brass, aluminium bronze, cupro–nickel alloys and stainless steels without excessive damage of such metals occurring.

When titanium is used in conjunction with other metals, it is essential that there is a favourable anode-to-cathode area ratio, i.e. that the surface area of titanium is as small as possible, while the surface area of the other metal, which acts as the anode, is large. Corrosion can also be prevented by the use of a sacrificial anode such as zinc in close proximity to the anodic metal.

Titanium has excellent erosion resistance, because of the ability of the metal to repair its protective oxide film.

RESISTANCE TO NITRIC ACID

Titanium is totally immune to aqueous solutions of nitric acid up to their boiling-points, and also to nitric acid with a concentration exceeding 70% at room temperature. Titanium is, however, attacked by nitric acid with a concentration between 30 and 60% at high temperatures, although such an attack is inhibited completely by the presence of traces of silicon in either the liquid or the vapour phase.

Titanium can thus be used with all concentrations of nitric acid, provided the temperature of the acid remains below about 120°C. It is widely used for processes which involve boiling nitric acid.

Titanium must not be used with red fuming nitric acid.

RESISTANCE TO HALOGENS

Titanium corrodes in liquified fluorine but withstands gaseous fluorine up to about 100°C. Titanium has better resistance to chlorine gas and chlorine compounds than virtually any other construction material. Titanium heat exchangers are therefore used for cooling chlorine gas, and titanium pumps, valves and pipes can be used when wet chlorine gas is to be handled. Titanium is particularly useful for handling sodium hypochlorite solutions.

Liquid bromine readily attacks titanium as does dry bromine gas. Titanium has, however, good resistance to moist gaseous bromine and bromine water.

Iodine vapour attacks titanium strongly at elevated temperatures, but corrosion caused by both dry and wet iodine vapour at room temperature is slight.

Chlorides have no effect on titanium at all at temperatures below about 100°C, although solutions of very high concentration of aluminium chloride, calcium chloride and aluminium chloride do affect titanium at elevated temperatures.

NON-OXIDISING ACIDS

The resistance of titanium to non-oxidising acids is not particularly good. Titanium withstands aqueous sulphuric acid up to a concentration of 5% at room temperature reasonably well, but at elevated temperatures even fairly dilute sulphuric acid solutions tend to attack titanium rapidly. Titanium has no resistance to hot concentrated sulphuric acid and corrosion is then fairly rapid. However, the corrosion rate in sulphuric acid is lowered drastically when chlorine is present. 60% H_2SO_4 attacks titanium rapidly at room temperature but has hardly any effect when saturated with chlorine gas (see Fig. 5.4).

The titanium–palladium alloy Titanium 260 has considerably more resistance to sulphuric acid than pure titanium, and can be used with a 4% H_2SO_4 solution at its boiling-point, or a 25% solution at room temperature.

The resistance of titanium to hydrochloric acid is poor. Ten per cent HCl has a corrosion effect of about 1 mm/year at room temperature and 7 mm/year at 60°C. Titanium cannot be used with hydrochloric acid stronger than this, even at room temperature. Again, the presence of free chlorine, nitric acid and other oxidising agents reduces the corrosion rate (see Fig. 5.5).

Phosphoric acid behaves somewhat similar to hydrochloric acid with respect to titanium. Corrosion rates are low in the cold but rise appreciably as temperatures are increased.

Titanium corrodes very rapidly in contact with even the most dilute hydrofluoric acid solutions.

The resistance of titanium to organic acids is excellent. The only exception to this is oxalic acid at high concentrations and temperatures.

RESISTANCE AGAINST ALKALIS AND OTHER MATERIALS

Titanium is totally inert against nearly all basic solutions at almost any temperature and any concentration. But boiling potassium hydroxide solution, exceeding 50% KOH, at 190°C corrodes titanium at about 2·5 mm/ year. Provided the temperature is kept below 130°C, it is possible to handle up to 70% NaOH solutions or 50% KOH solutions without appreciable corrosion occurring.

Titanium is virtually unattacked by salts, with the exception of sodium bisulphate.

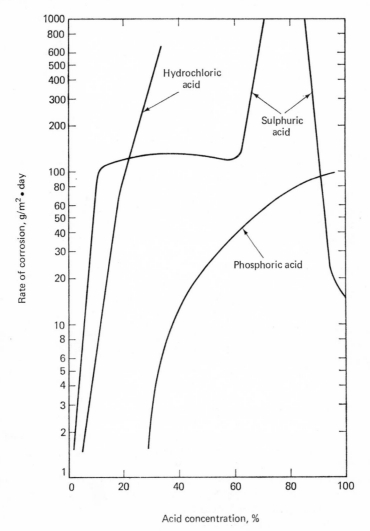

Fig. 5.4. Corrosion of titanium in inorganic acids at 25°C

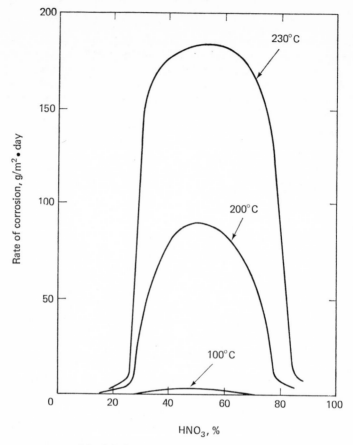

Fig. 5.5. Corrosion of titanium in HNO₃

The only organic materials that readily attack titanium are trichloroacetic acid and benzyl bromide at elevated temperatures. Against all other organic materials titanium is inert at all temperatures.

5.10 Tungsten

Tungsten has a density of 19·3 kg/dm³ and a melting-point of 3,370°C. It is not attacked by air even over long periods of time, nor by either hot or cold water. Its resistance to acids is excellent. At room temperatures tungsten is totally resistant to all dilute and concentrated acids, with the exception of a concentrated HF/HNO₃ mixture. At higher temperatures, concentrated acids have some corrosive effect upon tungsten, but in most cases corrosion is not too severe.

Tungsten resists hot alkali solutions extremely well and a feature is its complete immunity to liquid ammonia. Although tungsten does not normally react with acids, it is possible for it to be attacked under conditions where the tungsten structure is strongly anodic.

Tungsten reacts with fluorine gas at room temperature, with chlorine gas at temperatures above 260°C and with bromine and iodine at red heat.

When tungsten is heated in pure oxygen at a temperature of 600°C or higher, fairly rapid oxidation takes place on its surface. Molten alkalis dissolve tungsten rapidly, and molten oxidising salts such as nitrates, peroxides and chlorates attack the metal quite violently.

At high temperatures tungsten is also attacked by phosphorus, boron, carbon and silicon, as well as by some high temperature refractories such as MgO, Al_2O_3 and similar materials. The excellent corrosion resistance of tungsten is not always matched by its alloys. For example tungsten–iron alloys have poor resistance to acids even at room temperature.

5.11 Uranium

Uranium is a very heavy metal with a density of $18·7$ kg/dm^3 and a melting-point of 1,690°C. The $E°$ value of uranium is $+1·33$ V, which compares with $+0·76$ V for zinc and $+1·66$ V for aluminium. Uranium is therefore a fairly reactive metal which readily tarnishes in air to give UO_2 at atmospheric temperatures and U_3O_8 at temperatures above 200°C. At higher temperatures the oxide film is somewhat discontinuous and therefore oxidation proceeds rapidly.

The corrosion properties of uranium have been studied fairly comprehensively in recent years because of the vital importance of the metal in the field of nuclear energy. Because of the liability of uranium to corrode, it is quite impossible to have uranium in an unprotected form inside a nuclear reactor. In the low-temperature reactors the uranium is usually protected by the use of the magnesium alloy Magnox, and in high-temperature reactors either stainless steel or such zirconium alloys as ATR alloy or Zircaloy 2 are used for canning the uranium or its oxide.

Uranium resists cold water well, but only if oxygen is carefully excluded. Oxygenated cold water causes pitting corrosion of the metal. At 35°C the rate of corrosion is about $0·25$ g/m^2.day. As the temperature of the water increases, the rate of corrosion rises phenomenally. If uranium is put into contact with superheated water at 200°C, the amount of uranium which dissolves per square metre per *second* equals no less than 2 g. Fig. 5.6 shows the relationship between temperature and the rate of corrosion of uranium in contact with water.

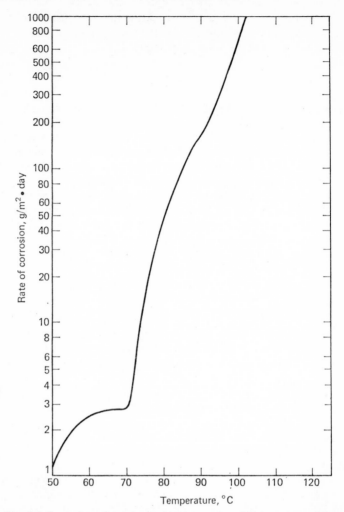

Fig. 5.6. Corrosion of uranium metal in water at different temperatures

Steam also attacks uranium badly. At a temperature below 450°C the oxide and hydride are formed; at higher temperatures water is broken down into uranium oxide and hydrogen gas.

Uranium is resistant to sulphuric acid up to 27% concentration in the cold, and has also good resistance to organic acids. It is readily attacked by even cold and dilute hydrofluoric, nitric, hydrobromic and hydriodic acid solutions and by fairly concentrated phosphoric and hydrochloric acids. It dissolves rapidly in cold and hot concentrated sulphuric acid.

The metal has a fairly good resistance to alkali hydroxides, but dissolves in ammonia solution.

The corrosion-resistant properties of uranium against aqueous solutions are very much enhanced by alloying the metal with 5–6% zirconium.

CORROSION RESISTANCE AGAINST OTHER AGENTS

Uranium is readily attacked by air and even inert gases such as argon and helium. Carbon dioxide has a very slight corrosive effect on uranium, provided that the temperature does not rise much above 200°C.

The corrosion rates quoted at 200°C are:

Air	112 g/m².day
Carbon dioxide	1·3 g/m².day
Argon	40 g/m².day
Helium	42 g/m².day

At elevated temperatures the corrosive attack by helium on uranium is very bad. Corrosion rates of 220 g/m².day have been quoted for helium gas at 310°C.

Uranium has a reasonable resistance to liquid sodium, particularly if traces of calcium metal are present, which inhibit attack. If oxygen is present molten sodium attacks uranium quite vigorously.

Uranium is attacked particularly rapidly by oxidising agents under alkaline conditions. Uranium corrodes quickly when sodium hydroxide, mixed with sodium peroxide, is brought into contact with the metal, and rapid attack is also induced by mixtures of ammonia and sodium persulphate.

5.12 Vanadium

Vanadium has a density of 5·87 kg/dm³ and a melting-point of 1,720°C. At the present moment vanadium is not widely used in industry, and where it is objects made from it are only small.

The metal has good resistance to aerated 60% sulphuric acid and 20% hydrochloric acid but can only withstand very dilute nitric acid solutions. At room temperature it is attacked slowly by 12% nitric acid, and rapidly by 17% nitric acid. Its resistance to phosphoric acid is also very poor, and the metal has no resistance to alkalis, which attack it rapidly. Corrosion in sea-water is about 1 g/m².day, and vanadium is also attacked by some salts, particularly sodium bisulphate and ferric chloride.

Vanadium has a good high-temperature resistance to air up to about 600°C, but above about 700°C the rate of surface oxidation increases rapidly.

The metal has poor resistance to molten metals.

5.13 Zinc

Zinc is inherently an active metal, and anodic to iron and steel, as well as many other metals. It is fairly corrosion-resistant to air, water and soils because it is covered quite rapidly with layers of protective compounds.

These are normally about 0·01 mm thick, and consist mainly of zinc oxide which is rapidly changed to basic zinc carbonate in air or water. In soils the protective layer is usually basic zinc sulphate.

CORROSION IN WATER

The rate of corrosion of zinc in distilled water (see Fig. 5.7) is somewhat higher than in hard waters, due to the fact that basic films are not readily formed. However, the main feature of corrosion of zinc in water is its marked dependence on temperature. Below 55°C corrosion rates in water are only very slight, due to the fact that a coherent oxide is formed. Once a temperature of 55°C is exceeded, the oxide film is no longer firm but becomes flaky. In consequence, the rate of corrosion rises very rapidly to 55 g/m².day at 60°C, at which temperature the rate of corrosion is a maximum. As the temperature is increased, the rate of corrosion falls again very rapidly to reach a very low value at about 90°C. The reason is that the oxide film once more has a firm and coherent character.

Zinc should therefore never be used in contact with water within the temperature range between 55 and 90°C.

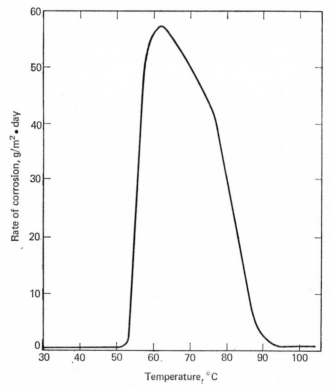

Fig. 5.7. Corrosion of zinc in distilled water at varying temperatures

Aeration of water increases the rate of attack, particularly if traces of carbon dioxide are present as well. For example, air agitation increases the rate of attack of zinc by distilled water at 30°C from 8 g/m^2.day to about 20 g/m^2.day. When zinc is submerged in water, it is also prone to pitting corrosion. The depth of the deepest pits can be as much as 10 times the average depth of corrosion.

The rate of corrosion of zinc in water, even within the critical temperature limits, is reduced drastically by the addition of certain inhibitors to the water. The most common ones used are: sodium dichromate, borax, sodium silicate or sodium hexametaphosphate.

CORROSION IN AQUEOUS SOLUTIONS

Zinc is badly affected by alkalis and acids. Corrosion rates are rapid below pH 6 and above pH 11.

Sodium chloride solutions tend to attack zinc somewhat, but the rate of attack varies considerably with both concentration and temperature. Once again the marked increase of corrosion rate within the critical temperature range of 55 to 90°C is clearly obvious (see Fig. 5.8).

The corrosion rate induced by chloride solutions also varies with the nature of the cation. Table 5.3 gives the rate of corrosion of zinc surfaces by 0·1N solutions at 25°C.

TABLE 5.3

0·1N solution in contact with zinc	Rate of corrosion at 25°C, g/m^2.day
NaCl	8·7
KCl	9·2
$CaCl_2$	3·2
NH_4Cl	3·8

High concentrations of zinc salts are poisonous, and for this reason zinc must not be used in contact with foodstuffs. In general, concentrations of zinc salts in aqueous liquids below about 40 ppm are harmless. It is therefore quite satisfactory to keep cold water in galvanised pipes or galvanised tanks, as final concentrations of zinc ions in the water after even very prolonged periods of exposure never reach anywhere near the above figure. Zinc, however, dissolves rapidly in milk or carbonated acidic beverages and danger from poisoning could then become very real.

ATMOSPHERIC CORROSION

Zinc sheeting exposed to the atmosphere corrodes very slowly, particularly if the atmosphere is not heavily polluted. The following figures of the rate of corrosion of zinc roofing in varying atmospheres usually apply:

I

Very dry hot area (desert)	Less than 0·005 g/m².day
Rural area in temperate region	0·02 g/m².day
Marine area in temperate region	0·04 g/m².day
Marine and industrial area	0·12 g/m².day
Heavily industrialised area	0·15 g/m².day

1 g/m².day corresponds to an average corrosion rate of about 0·05 mm/year.

ZINC–IRON COUPLES

In most chemical solutions the rate of corrosion of zinc is about the same order of magnitude as the corrosion of iron and steel. Exceptions are water, where the rate of corrosion of zinc is far less. If steel and zinc are coupled together, the corrosion rate of steel falls virtually to zero, and zinc dissolves preferentially. This is the basis of the use of zinc as a sacrificial anode.

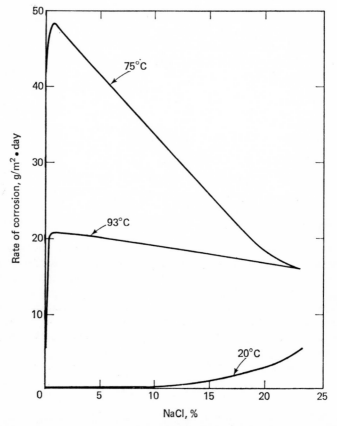

Fig. 5.8. Corrosion of zinc in sodium chloride solutions of varying strengths and at different temperatures

Table 5.5 gives corrosion rates of iron and zinc, when exposed separately to different solutions and when coupled together.

TABLE 5.5

Nature of solution	Corrosion rate at 25°C, mm/year			
	Iron alone	Zinc alone	Zinc–iron couple	
			Iron	Zinc
0·1N MgSO$_4$	0·065	nil	nil	0·085
0·1N Na$_2$SO$_4$	0·25	0·27	nil	0·88
0·1N NaCl	0·25	0·25	nil	0·75
0·01N NaCl	0·18	0·11	nil	0·22
H$_2$CO$_3$ (sat.)	0·072	0·01	nil	0·037
Ca(HCO$_3$)$_2$	0·15	nil	nil	nil
tap water	0·07	nil	nil	nil

Zinc coatings on steel have the property of very much reducing liability to fatigue fractures. In one experiment performed, a steel bar was vibrated for 10^7 cycles in a corrosive atmosphere and at the end of that time its ultimate tensile strength was equal to 210 MN/m². A similar steel bar, coated with a thin surface layer of zinc was vibrated in the same way in the same atmosphere, and was able to withstand 420 MN/m².

CORROSION OF ZINC WITH ORGANIC LIQUIDS

Zinc has excellent corrosion resistance to nearly all neutral organic materials, provided these are free from water. Zinc is particularly satisfactory to use in conjunction with petrol, oils, glycerol, trichlorethylene and other chloro compounds. If traces of water are present, rapid local corrosion can occur.

5.14 Zirconium

Zirconium is a corrosion-resistant metal with properties rather similar to titanium. Its density is quite low, namely 6·45 kg/dm³, and its melting-point is 1,852°C.

Its use has expanded enormously in recent years, particularly within the nuclear energy field. It is mainly used in the reactor field in the form of the following two alloys:

Zircaloy 2: 1·5% Sn, 0·1% Cr, 0·12% Fe, 0·05% Ni, rest is hafnium-free zirconium

ATR alloy: 0·5% Cu, 0·5% Mo, rest is zirconium with traces of hafnium

Zircaloy 2 is used in conjunction with water-cooled reactors such as the boiling water reactor, the D$_2$O moderated reactor and the pressurised water reactors. The ATR alloy is used for gas-cooled reactors. The main purpose of these alloys in the nuclear power field is for canning the fuel elements.

RESISTANCE AGAINST ACIDS AND ALKALIS

Zirconium is not as corrosion-resistant as titanium in oxidising media, but it is better under non-oxidising conditions and against alkalis. As is usual, the corrosion resistance of zirconium against acids reduces as the strength of the acid concerned increases and as the temperature goes up.

It is considered that the corrosion rate of zirconium is within tolerable limits provided the following concentrations and temperatures are not exceeded:

Hydrochloric acid	Up to 20%	Up to 100°C
Phosphoric acid	Up to 30%	Up to 100°C
Sulphuric acid	Up to 70%	Up to 100°C
Nitric acid	Up to 70%	Up to 200°C
Organic acids	Any strength	Any temperature
Hydrofluoric acid	Very poor resistance at all strengths and any temperature	
Aqua regia	Readily attacked in the cold	

Zirconium has excellent resistance to aqueous alkalis at all strengths and temperatures although it is very slightly attacked by fused alkalis. Its resistance to all forms of water and aqueous salt solutions is excellent, with the exception that it is attacked by hot and concentrated solutions of strongly oxidising salts such as potassium dichromate, potassium chlorate and similar.

Zirconium is therefore used as a lining in chemical plants where good resistance to hot concentrated acids and alkalis is required. It should, however, be mentioned that the fine corrosion resistance of zirconium can be destroyed if tiny quantities of carbon are present in the metal. The limit of carbon permitted is as low as 0·06%. Any carbon in excess of this can cause corrosion rates 100 times larger than those that take place with the unadulterated metal.

Literature sources and suggested further reading

1   LA QUE, F. L., and COPSON, H. R., *Corrosion resistance of metals and alloys*, Reinhold, New York (1963)
2   *Corrosion resistance of titanium*, ICI Limited, Birmingham (1969)
3   GREATHOUSE, G. A., and WESSEL, C. J., *Deterioration of materials*, Reinhold, New York (1954)
4   'Corrosion chart', *Chemical Processing* (December 1969)
5   MOREX LIMITED, Brochures dealing with molybdenum, niobium, tantalum, tungsten and zirconium.

6 ZINC DEVELOPMENT ASSOCIATION, *The performance of zinc in buildings*
7 MATHEWSON, C. H., *Zinc, the metal, its alloys and compounds*, Reinhold, New York (1959)
8 GODDARD, H. P., et al., *The corrosion of light alloys*, John Wiley, New York (1967)

Chapter 6   Corrosion-resistant Alloys

6.1 Types of stainless steel

The fundamental basis underlying stainless steel is the fact that when steel that contains a high percentage of chromium, is exposed to an oxidising atmosphere, an adherent film of chromic oxide (Cr_2O_3) is formed. This acts as a passive surface, preventing the corrosion of the metal beneath. Even if such a film is broken, it reforms provided oxygen is present. Stainless steel can, however, fail if the chromium oxide film is broken in the absence of oxygen. The higher the percentage of chromium in the alloy, the better the protective action on the steel, and the addition of nickel serves to improve the stability of the oxide film. There are almost one hundred different types of stainless steel on the market; these tend to fall into the following three groups:

1 Martensitic stainless steels.
2 Ferritic stainless steels.
3 Austenitic stainless steels.

MARTENSITIC STAINLESS STEELS

The martensitic stainless steels contain about 0·1–0·4% carbon and about 14% chromium. They are very hard and can be quenched and tempered like normal tool steels. Other elements such as manganese and nickel are added to stabilise their structures, but this is by no means universal. A martensitic stainless steel with 0·35% carbon can have a tensile strength as high as 850 MN/m^2 and a Brinell hardness up to 230 and would be used for the manufacture of stainless steel cutting tools. When the carbon content is somewhat lower the material is not quite as hard, but brittleness is reduced. Such forms of martensitic stainless steel are used in the manufacture of wires, strands and cables, jointing brackets, etc. Excessive cold-working reduces the corrosion resistance of martensitic stainless steels which are far less corrosion-resistant than either ferritic or austenitic steels. Under the action of many chemical reagents they become liable to inter-granular attack and stress cracking.

FERRITIC STAINLESS STEELS

These are very soft stainless steels with a low tensile strength but a high elongation figure. They contain a maximum of 0·1% carbon and between

15 and 20% chromium. If the percentage carbon content of ferritic stainless steels exceeds 0.1%, ductility is severely affected. The tensile strength of normal ferritic stainless steel is about 400–500 MN/m² and elongation on a 50 mm long sample is up to 40%. Impact resistance is, however, poor, with Izod impact figures as low as 1–2 kg.m being quoted. Ferritic stainless steel is used almost exclusively for the production of pressed and deep-drawn articles such as kitchen sinks, bowls and similar products. Their corrosion resistance does not differ unduly from that of most grades of austenitic stainless steel.

AUSTENITIC STAINLESS STEELS

The austenitic stainless steels (see Fig. 6.1) are by far the most widely used. Originally they were based upon an approximate formulation of

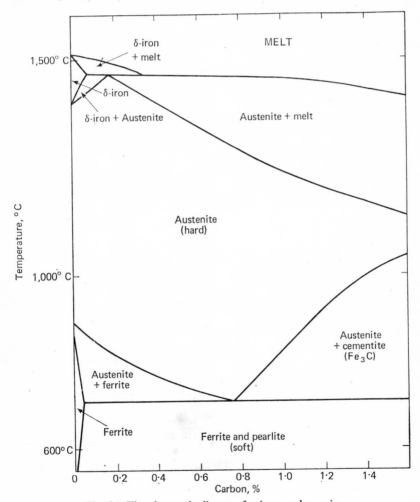

Fig. 6.1. The phase rule diagram for iron–carbon mixes

18% chromium and 8% nickel and such an alloy is still known as an 18–8 stainless steel. The austenitic structure, which is a solid solution of iron carbide and all the metal additives in iron, is stabilised at all temperatures, producing a soft and flexible product. During cold-working some of the austenite is converted to martensite and this hardens the steel and increases its strength. Austenitic stainless steels with a carbon content of up to 1% have tensile strengths of around 750 MN/m² and yield strengths of about 270 MN/m². The elongation varies between 30 and 73% depending on carbon content, and the Brinell hardness between 160 and 200. The impact strength is around 15–18 kg.m (Izod impact).

Tables 6.1 and 6.2, which are produced here by permission of the International Nickel Company, give the main wrought and cast chromium–

TABLE 6.1 AISI designations

AISI type	Composition, %								
	C max	Mn max	P max	S max	Si max	Cr	Ni	Mo	Other
201	0·15	5·50–7·50	0·060	0·030	1·00	16·00–18·00	3·50–5·50	—	N 0·25 max
202	0·15	7·50–10·00	0·060	0·030	1·00	17·00–19·00	4·00–6·00	—	N 0·25 max
301	0·15	2·00	0·045	0·030	1·00	16·00–18·00	6·00–8·00	—	—
302	0·15	2·00	0·045	0·030	1·00	17·00–19·00	8·00–10·00	—	—
302B	0·15	2·00	0·045	0·030	2·00–3·00	17·00–19·00	8·00–10·00	—	—
303	0·15	2·00	0·20	0·15 min	1·00	17·00–19·00	8·00–10·00	0·60 max	—
303Se	0·15	2·00	0·20	0·06	1·00	17·00–19·00	8·00–10·00	—	Se 0·15 min
304	0·08	2·00	0·045	0·030	1·00	18·00–20·00	8·00–12·00	—	—
304L	0·03	2·00	0·045	0·030	1·00	18·00–20·00	8·00–12·00	—	—
305	0·12	2·00	0·045	0·030	1·00	17·00–19·00	10·00–13·00	—	—
308	0·08	2·00	0·045	0·030	1·00	19·00–21·00	10·00–12·00	—	—
309	0·20	2·00	0·045	0·030	1·00	22·00–24·00	12·00–15·00	—	—
309S	0·08	2·00	0·045	0·030	1·00	22·00–24·00	12·00–15·00	—	—
310	0·25	2·00	0·045	0·030	1·50	24·00–26·00	19·00–22·00	—	—
310S	0·08	2·00	0·045	0·030	1·50	24·00–26·00	19·00–22·00	—	—
314	0·25	2·00	0·045	0·030	1·50–3·00	23·00–26·00	19·00–22·00	—	—
316	0·08	2·00	0·045	0·030	1·00	16·00–18·00	10·00–14·00	2·00–3·00	—
316L	0·03	2·00	0·045	0·030	1·00	16·00–18·00	10·00–14·00	2·00–3·00	—
317	0·08	2·00	0·045	0·030	1·00	18·00–20·00	11·00–15·00	3·00–4·00	—
D319	0·07	2·00	0·045	0·030	1·00	17·50–19·50	11·00–15·00	2·25–3·00	—
321	0·08	2·00	0·045	0·030	1·00	17·00–19·00	9·00–12·00	—	Ti 5×C min
347	0·08	2·00	0·045	0·030	1·00	17·00–19·00	9·00–13·00	—	Nb Ta 10×C min
348	0·08	2·00	0·045	0·030	1·00	17·00–19·00	9·00–13·00	—	Nb Ta 10×C max; Co 0·20 max

nickel stainless steels as well as their American Institute of Iron and Steel (AISI) and Alloy Casting Institute (ACI) reference numbers. (My thanks are also due to the International Nickel Company for providing me with most of the information on which this chapter is based.)

CORROSION RESISTANCE OF AUSTENITIC STAINLESS STEELS

The austenitic stainless steels, like other stainless steels, are protected against corrosion by virtue of the formation of an impervious film on their surfaces. In general, one aims at less than about 2·5 μm/year corrosion. If a sample shows a resistance to attack of this magnitude or less, it is considered to be satisfactory for the job in hand.

TABLE 6.2   Alloy Casting Institute classification of chromium–nickel stainless steel castings

Cast alloy designation	Wrought alloy type*	Composition %								
		C max	Mn max	P max	S max	Si max	Cr	Ni	Mo	Other
CD–4MCu	—	0·040	1·00	0·04	0·04	1·00	25–27	4·75–6·00	1·75–2·25	Cu 2·75–3·25
CE–30	—	0·30	1·50	0·04	0·04	2·00	26–30	8–11	—	—
CF–3	304L	0·03	1·50	0·04	0·04	2·00	17–21	8–12	—	—
CF–8	304	0·08	1·50	0·04	0·04	2·00	18–21	8–11	—	—
CF–20	302	0·20	1·50	0·04	0·04	2.00	18–21	8–11	—	—
CF–3M	316L	0·03	1·50	0·04	0·04	1·50	17–21	9–13	2·0–3·0	—
CF–8M	316	0·08	1·50	0·04	0·04	1·50	17–21	9–12	2·0–3·0	—
CF–12M	316	0·12	1·50	0·04	0·04	1·50	18–21	9–12	2·0–3·0	—
CF–8C	347	0·08	1·50	0·04	0·04	2·00	18–21	9–12	—	Nb 8 × C min, 1·0 max or Nb Ta 10 × C min
CF–16F	303	0·16	1·50	0·17	0·04	2·00	18–21	9–12	1·5 max	Se 0·20–0·35
CG–8M	317	0·08	1·50	0·04	0·04	1·50	18–21	9–13	3·0–4·0	—
CH–20	309	0·20	1·50	0·04	0·04	2·00	22–26	12–15	—	—
CK–20	310	0·20	1·50	0·04	0·04	2·00	23–27	19–22	—	—
CN–7M	—	0·07	1·50	0·04	0·04	1·50	18–22	27–31	1·75–2·50	Cu 3·0 min

*Wrought alloy type numbers are included only for the convenience of those who wish to determine corresponding wrought and cast grades. The chemical composition ranges of the wrought materials differ from those of the cast grades.

6.2 Corrosion of stainless steels in various environments

MARINE CORROSION

The oxygen content of normal sea-water ranges from 6–10 ppm between temperatures of 4 and 32°C. This is adequate to maintain the passive film on stainless steels when the water flow over the surfaces exceeds 1·5 m/sec. If the flow velocity is lower, it is possible for debris, silt and fouling organisms to collect on the surface of the stainless steel and to establish crevices which limit or even completely eliminate the supply of oxygen to the surfaces. Pitting corrosion can then take place but it has been found that stainless steels which contain molybdenum have considerable resistance to stagnant sea-water and that pitting is far less under such circumstances. The worst performance of all austenitic stainless steels in stagnant sea-water was found with grade AISI 308. This exhibited maximum pit depths of 5·2 mm and average pit depths of 2·1 mm, with crevice corrosion depths of up to 3·6 mm after a year's exposure. In contrast alloy 316 which contains 3·18% molybdenum has a maximum pit depth of 0·64 mm and an average pit depth of 0·28 mm after a year's exposure to stagnant sea-water.

When stainless steel is exposed to alternate wetting and drying in sea-water there is an opportunity for passive films to be formed, and in consequence pitting is much reduced. In the case of a sample of 316 stainless steel the depth of pits found with alternative wetting and drying over a year was 0·36 mm, as against 0·94 mm with total immersion.

When sea-water is moving at high velocities past stainless steel surfaces virtually no corrosion takes place, even under conditions which may cause the perforation of 1·6 mm thick medium carbon steel sheets due to cavitation erosion.

Stainless steels in the passive or oxidised conditions have very noble potentials in sea-water. If, however, the passive film is destroyed in places, the potential is very much higher.

TABLE 6.3   Corrosion of medium carbon steel in sea-water when in contact with different metals
The temperature is 10°C and the flow-rate 2·4 m/sec. Areas of steel and coupled metal are equal

Couple	Corrosion rate, mm/year	Galvanic effect, mm/year
Steel (alone)	0·79	—
Steel/304 stainless steel	0·91	0·12
Steel/titanium	1·07	0·28
Steel/copper	2·54	1·75

For example, the difference in potential between active and passive 316 stainless steel is 0·13 V and between active and passive 304 stainless steel is 0·45 V. For this reason heavy corrosion of active stainless steel is to be

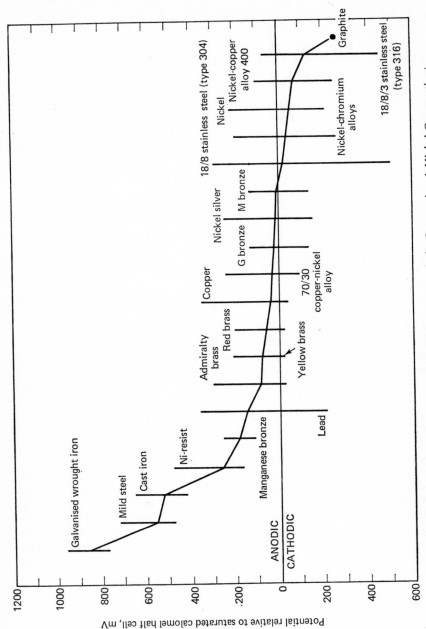

Fig. 6.2. The galvanic series in sea-water (By courtesy of the International Nickel Corporation)

expected whenever small areas of active material are coupled to large areas of passive stainless steel. Stainless steel, in spite of its noble character, does not unduly increase the rate of corrosion of medium carbon steel which is in contact with it. In fact, its effect is far less than that of metals such as copper and titanium. This can be seen clearly from the Table 6.3. Where the area of stainless steel is small in contrast with normal steel, no increase in the rate of corrosion of the normal steel can be observed. On the other hand, the rate of corrosion of stainless steel, even in stagnant sea-water, is very much reduced. Austenitic stainless steel is widely used in sea-water for applications where it is bonded to relatively large areas of carbon steel, such as ships hulls.

CHEMICAL ENVIRONMENTS

Stainless steels have excellent resistance to atmospheric exposure, and retain their attractive appearance for a long time even in highly polluted atmospheres. Neither cold-forming nor cold-rolling reduces the corrosion resistance. As far as other chemical environments are concerned, stainless steels perform best in oxidising conditions.

INTERGRANULAR CORROSION

When austenitic stainless steels are held between 425 and 900°C they may undergo a change which makes them susceptible to intergranular corrosion by a number of otherwise quite harmless reagents. The reason for this is that chromium carbides tend to precipitate at grain boundaries, thereby denuding adjacent areas of chromium. This can happen during welding operations, but the original state can be restored by annealing at 1,000 to 1,120°C, followed by rapid cooling through the sensitive range. Stainless steels in which the carbon content is held below 0·03%, as in the case of types 304 L and 316 L, avoid sensitisation (becoming liable to stress corrosion) during welding. The carbon content must be even lower (0·02% or less) if complete immunity from sensitisation is to be obtained if the stainless steel is held within the critical temperature range. Stainless steels which have small quantities of niobium and titanium in their formulations, such as types 321 and 347, are recommended for use in corrosive environments which may otherwise cause intergranular attack near welds. Of the two, niobium is better since titanium at welding temperatures tends to react with gases.

STRESS CORROSION CRACKING

Stainless steels which retain residual stresses may develop stress corrosion cracks in certain chemical solutions, particularly chlorides. The addition of elements such as niobium and titanium seems to have little effect. Annealing at temperatures in excess of about 900°C is the only answer.

PITTING AND CREVICE CORROSION

Corrosion develops at places where the passivity has been destroyed, and these areas then become anodic with respect to the remaining surface of the stainless steel. Because the passive area is inevitably larger than the active one, the cathodic reaction is the rate-determining one and the rate of pitting which takes place at the active area is controlled by the nature of the depolarisers. Reagents which cause rapid pitting are: oxygen in stagnant solutions, ferric chloride, hypochlorous acid and mercuric chloride. Nitrates and chromates, on the other hand, are extremely effective in stopping the formation of active areas on stainless steel and thus prevent corrosion.

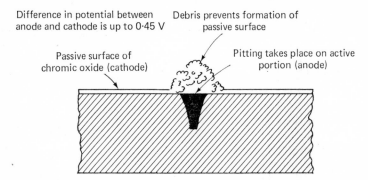

Fig. 6.3. Mechanism of corrosion on stainless steel where portions are covered by debris

The role of dissolved oxygen is rather interesting. If there is no oxygen present in solution, there is no depolarisation of the cathodic areas and therefore no corrosion. If oxygen is present is small quantities only, pitting corrosion takes place quite rapidly. Although there is insufficient oxygen to passivate the active patches, there is more than enough to enable the depolarisation action to take place. Liberal aeration, on the other hand, produces a passive film all over and stops pitting. Any debris adhering to stainless steel stops oxygen from getting to the surface and thus induces pitting at the affected parts. The addition of 2–3% of molybdenum to high nickel stainless steel considerably reduces corrosion in such media as sulphuric acid, sulphurous acid, phosphoric acid, formic and other organic acids.

When designing stainless steel equipment, all shapes and joints which include crevices or deep recesses must be avoided, as otherwise oxygen cannot get at the metal and in consequence the crevices become anodes.

RESISTANCE OF STAINLESS STEEL TO WATER AND SALT SOLUTIONS

Normal austenitic stainless steels show virtually complete resistance to waters of all kinds even at high temperatures and pressures. Stainless steel is hardly attacked, even under conditions that cause very heavy corrosion on normal steel structures. Similarly, the austenitic stainless steels resist corrosion by neutral and alkaline salt solutions, including those of a strongly oxidising nature.

Stainless steel is less satisfactory when in contact with halogen salts. Pitting corrosion can then easily take place. Acid salts in oxidising solutions are particularly bad and should not be used in stainless steel equipment. Often pitting becomes pronounced at certain critical temperatures. In contact with 3N sodium chloride solution, the critical temperatures are about 60°C for 304 steel and 75°C for 316 steel. Yet at about 90°C the rate of pitting is lower, owing to the fact that the amount of oxygen dissolved falls off at elevated temperatures.

RESISTANCE OF STAINLESS STEEL TO ACIDS

Stainless steel must never be used in contact with hydrochloric acid, as this rapidly destroys it even in the cold. In the hot the metal reacts quickly with the evolution of hydrogen. Even very low concentrations have an adverse effect. Type 316 stainless steel resists sulphuric acid solutions below 20% H_2SO_4 and above 90% H_2SO_4 at room temperature (see Fig. 6.4). Stainless steel cannot be used in contact with sulphuric acid between these two concentration limits even at room temperature. The corrosion rate of stainless steel in contact with sulphuric acid increases as the temperature is raised, but is markedly reduced if the acid contains oxidising agents.

Ferric sulphate, copper sulphate, nitric acid, chromic acid and dissolved oxygen all tend to reduce the rate at which sulphuric acid corrodes stainless steel. For example, one can use stainless steel vessels for handling mixed nitric and sulphuric acids in nitration plants. Stainless steels which contain small quantities of molybdenum and copper, as well as traces of silicon have better resistance to sulphuric acid than the basic 18–8 types. A stainless steel which is used specially for handling up to 50% and above 90% sulphuric acid at temperatures up to 80°C is one containing 24% Ni, 20% Cr, 3% Mo, 3·5% Si, 1·75% Cu, 0·6% Mn and less than 0·05% C. This alloy can be used with 65% H_2SO_4 up to 60°C and for other concentrations up to about 50°C.

Stainless steel has excellent resistance to nitric acid at all concentrations and at virtually all temperatures. Up to the boiling-point of the acid types 304 and 347 are useful, but for extreme temperature conditions types 309 and 310, containing some niobium, are better because they have resistance to intergranular attack, which may occur at welds.

Phosphoric acid too has hardly any effect on stainless steel, type 316 being the best to use at elevated temperatures. Sulphurous acid causes pitting on types 302 and 304, but has no effect on stainless steels that contain molybdenum. Types 316 and 317 are used for handling sulphite liquors in the paper industry.

CORROSION OF STAINLESS STEEL BY OTHER SUBSTANCES

The austenitic stainless steels have excellent resistance to weak bases, such as ammonium hydroxide, and organic bases such as aniline, pyridine, etc.

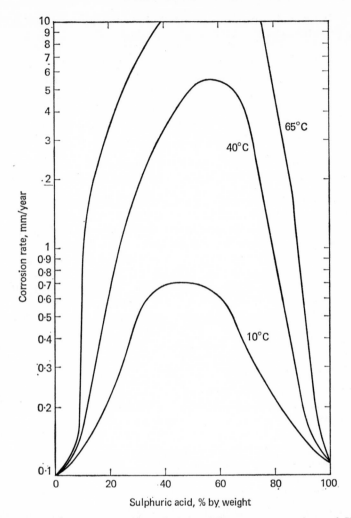

Fig. 6.4. Corrosion of 316 stainless steel by different concentrations of H_2SO_4 at varying temperatures

Up to a concentration of 50% and a temperature of 100°C stainless steel is also not unduly affected by sodium and potassium hydroxides. Stainless steel should not be used if conditions are more drastic than these, as corrosion is then quite appreciable. Caustic alkalis may also induce stress corrosion cracking. Stainless steels are ideal for handling organic compounds. The only exceptions are some organic acids such as formic and lactic acids, which can only be handled with types 316 and 317, and some organic halides. These last hydrolyse with water and then liberate hydrochloric acid which, as has already been pointed out, damages stainless steel severely.

Foods that contain organic acids should be handled in 304 and 316 stainless steels, whereas those containing salt should only be handled in the latter.

Dry and cold chlorine gas (see Fig. 6.5) can be handled in 304, 316 and 317 stainless steels, but stainless steel should never be used for wet or hot chlorine gas. Stainless steel is also not really suitable for handling chlorine-containing liquids such as sodium hypochlorite, although it is possible to keep solutions containing less than 0·3% of available chlorine in stainless steel plant made from types 316 or 317.

HIGH-TEMPERATURE CORROSION OF AUSTENITIC STAINLESS STEEL

Stainless steels are used increasingly for high-temperature conditions. It has been found that the resistance of stainless steel surfaces to gases is largely dependent upon the resistance of the surface coating formed to spalling effects, i.e. to break-up and detachment.

When stainless steels are exposed to air at 1,000°C the higher nickel alloys perform best because nickel oxide has a lower thermal expansion figure than chromium oxide. Types 310, 330 and particularly alloy 800, which contains 32% Ni and 21% Cr, are the most suitable.

If the air is contaminated by other gases the corrosion rate rises manifold.

Table 6.4 gives data for 18–8 stainless steel at 900°C over a period of 500 hr.

TABLE 6.4

Type of atmosphere	Weight gain, g/m²
Dry air	50
Air + 5% water vapour	350
Air + 5% water vapour + 5% SO_2	400
Air + 5% water vapour + 5% CO_2	500

Austenitic stainless steels have excellent resistance to steam up to about 900°C. Corrosion rates for all varieties of stainless steel seem to be negligible even under conditions of prolonged exposure.

Stainless steel is attacked at elevated temperatures by sulphur dioxide, hydrogen sulphide and sulphur vapour, of which the latter two are the worst. The rate of attack found with 316 stainless steel containing niobium, exposed to a mixture of air and sulphur dioxide at 650°C was 0·013 mm/ year but a concentration of only 0·1% of hydrogen sulphide by volume in a hydrogen atmosphere at between 12 and 35 bar causes a corrosion rate of 0·5 mm/year at 650°C.

The most successful stainless steel alloys to be used in atmospheres which contain high proportions of either hydrogen sulphide or sulphur vapour at high temperatures are types 309, 310 and 314, whilst 316 and 321 are not satisfactory.

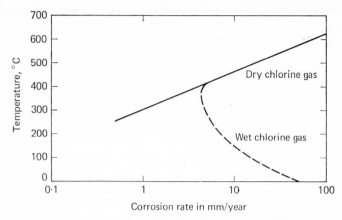

Fig. 6.5. Corrosion of 18–18 stainless steel by chlorine

RESISTANCE TO FLUE GASES

Flue gases can be either oxidising or reducing. Oxidising flue gases contain an excess of oxygen and sulphur in the form of sulphur dioxide. Reducing flue gases contain hydrogen and carbon monoxide and sulphur in the form of hydrogen sulphide. Both types of flue gas contain carbon dioxide and nitrogen which are, however, relatively inert under high-temperature conditions. Stainless steel is attacked by oxidising flue gas at a rate which is somewhat higher than the rate of attack in air. As the sulphur content is increased the corrosion rate also increases. Reducing flue gases have a roughly similar effect to oxidising flue gases when the sulphur content is less than about 2·3 g/m³. As the sulphur content increases, the rate of corrosion rises much more rapidly than it does with oxidising flue gases.

The rate of corrosion of stainless steels by both oxidising and reducing flue gases is markedly dependent upon the percentage chromium content of the stainless steel and, to a smaller extent, upon the percentage nickel content. Table 6.5 gives the average corrosion rates of a number of different stainless steel alloys under oxidising and reducing conditions, when they are exposed to a temperature of 1,000°C, and when the sulphur content is equal to 2·3 g/m³.

Stainless steels have good resistance at all temperatures and pressures to attack by hydrogen which normally can diffuse through steel. Austenitic stainless steels have been found satisfactory for use in ammonia converters, where the approximate working composition of the gases present was 60% H_2, 20% N_2, 8% Ar and 12% NH_3. The corrosion rates at 550°C varied between 2·5 and 18μm, depending on grade, the best results being achieved by grades 310, 314 and 330. Stainless steel is most unsatisfactory for use with pure ammonia at high temperatures. Corrosion rates at 500°C vary from 1·4 mm/year with type 310 to 13·2 mm/year for type 316.

As mentioned previously, stainless steel is unsatisfactory for use with hot chlorine or hydrogen chloride gas and is in fact even worse with respect

K

to fluorine. Niobium-modified 309 and 310 stainless steels can handle dry fluorine up to a temperature of 250°C, but are severely attacked if the temperature rises beyond this.

TABLE 6.5

Alloy		Corrosion rate, mm/year	
Cr, %	Ni, %	Oxidising conditions	Reducing conditions
17	0	25·0	60·0
20	0	5·0	25·0
25	0	1·2	1·2
30	0	0·5	Nil
20	10	1·8	3·5
20	15	1·2	2·5
20	20	0·8	1·2
20	30	0·4	0·5

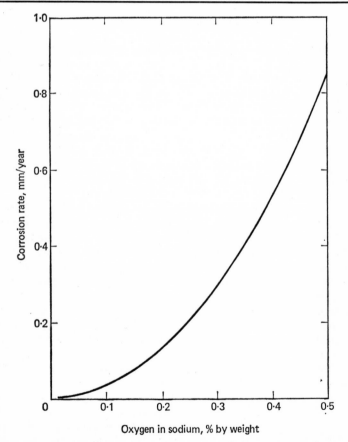

Fig. 6.6. Corrosion of 347 stainless steel by liquid sodium containing traces of oxygen at 700°C

REACTIONS WITH MOLTEN METALS

Stainless steels can be used to handle molten sodium and sodium/potassium alloys, but care must be taken that these are not adulterated by even traces of oxygen or carbon (see Fig. 6.6). Use is made of this important property of the 18–8 stainless steels in the design of heat exchangers for fast or breeder nuclear reactors. Stainless steel has poor resistance to molten lead and is also not too satisfactory for handling most other molten metals, with the exception of lithium and thallium.

ATTACK BY VANADIUM PENTOXIDE

Vanadium pentoxide is produced during the combustion of many oils. Fuels containing as little as 50 ppm of vanadium can cause severe attack on stainless steels at temperatures above 750°C (see Fig. 6.7). Only at temperatures below the melting-point of the ash is the corrosive effect of vana-

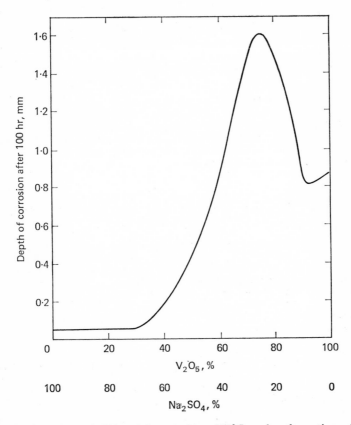

Fig. 6.7. Corrosion of 304 stainless steel at 900°C under the action of various Na$_2$SO$_4$/V$_2$O$_5$ deposits

dium pentoxide slight. Stainless steels do not therefore constitute an answer to vanadium pentoxide corrosion.

6.3 Nickel-based corrosion-resistant alloys

In addition to the stainless steels there are also a large number of corrosion-resistant alloys, which use mainly nickel, chromium and iron in their formulations. Table 6.6 (International Nickel Company) gives the main alloys of this kind made by the firm together with the trade names used. Obviously

TABLE 6.6   Common corrosion-resistant nickel alloys

Composition, %	Ni	Fe	Cr	Mo	Cu	C	Si	Mn	Other
Iron-base nickel–chromium–copper–molybdenum alloys									
Illium alloy P (*c*)	8	57	28	2·25	3·25	0·20	0·75 max	1·0 max	—
Worthite, stainless (*d*)	24	48	20	3	1·75	0·07 max	3·25	0·6	—
Carpenter Stainless No. 20 Cb	29	44	20	2·5	3·3	0·07 max	0·6	0·75	Nb(+Ta) 0·6
ACI Type CN–7M (*a*)	29	44	20	2·2	3·3	0·07 max	1·5 max	1·5 max	—
Carpenter Stainless No. 20 Cb–3	34	39	20	2·5	3·3	0·07 max	0·6	0·75	Nb(+Ta) 0·6
Nickel-base iron–chromium–molybdenum alloys									
Incoloy alloy 825	41·8	30	21·5	3·0	1·80	0·03	0·35	0·65	Al 0·15; Ti 0·90
Hastelloy alloy G (*b*)	45	19·5	22·2	6·5	2·0	0·03	0·35	1·3	W 0·5; Nb+ Ta 2·12
Hastelloy alloy F (*b*)	47	17	22	6·5	—	0·05 max	—	—	Co 2·5 max; W 0·6; Other 5·5
Nickel-base chromium–molybdenum alloys									
Illium alloy 98 (*c*)	55	1·0	28	8·5	5·5	0·05	0·7	1·25	—
Illium alloy G (*c*)	56	6·5	22·5	6·4	6·5	0·20	0·65	1·25	—
Inconel alloy 625	61	3	22	9·0	0·1	0·05	0·3	0·15	Nb 4
Illium alloy R	68	1	21	5	3	0·05	0·7	1·25	—
Nickel-base molybdenum–chromium alloy									
Hastelloy alloy C (*b*)	54	5	15·5	16	—	0·08 max	1·0 max	1·0 max	Co 2·5 max; W 4; V 0·4 max
Nickel-base molybdenum alloy									
Hastelloy alloy B (*b*)	61	5	1 max	28	—	0·05 max	—	—	Co 2·5 max; Other 3

Composition, %	Ni	Fe	Cr	Mo	Cu	C	Si	Mn	Other
Nickel-base silicon alloy									
Hastelloy alloy									
D (*c*)	82	2 max	1 max	—	3·0	0·12 max	9	—	Co 1·5 max; Other 2
Nickel-alloyed cast irons									
Ni-resist Type 1 (*c*)	15‘5	69	2·5	—	6·5	2·8	2·0	1·2	—
Ni-resist Type 2 (*c*)	20	72	2·5	—	—	2·8	2·0	1·0	—
SG Ni-resist Type									
D–2 (*c*)	20	72	2	—	—	3·0 max	2	0·85	P 0·08 max
Nickel–copper alloys									
Monel alloy 400	66	1·35	—	—	31·5	0·12	0·15	0·90	—
Copper–nickel									
alloy CA 715	31	0·55	—	—	67	—	—	1·0 max	Pb 0·05 max; Zn 1·0 max.
Nickel									
Nickel 200	99·5	0·15	—	—	0·05	0·06	0·05	0·25	—
Nickel-base chromium alloys									
Corronel alloy 230	55 min	5·0 max	36	—	1·0 max	0·08 max	0·6 max	1·0 max	Ti 1·0 max; Al 0·5 max
Iconel alloy 600	76	7·2	15·8	—	0·10	0·04	0·20	0·20	—
Iron–nickel–chromium alloy									
Incoloy alloy 800	32	46	20·5	—	0·30	0·04	0·35	0·75	—

(*a*) Cast '20' alloys, such as Duriment 20, Aloyco 20, etc.
(*b*) Composition of wrought alloy; available also in cast form
(*c*) Available in cast form only
(*d*) Composition of cast alloy

other companies will use somewhat different formulations, with different trade names, and this list cannot be considered to cover all or even the majority of alloys made. It serves simply as an indication of the nature of modern corrosion-resistant alloys.

Nickel-based corrosion-resistant alloys of the type shown are mainly used for the following purposes:

1 For the construction of chemical plant.
2 For use in marine environments.
3 To avoid fuel ash corrosion in furnaces.

CHEMICAL PLANT CONSTRUCTION

The nickel alloys are far more effective in their resistance to phosphoric and many other acids than stainless steel. For example, the corrosion rate of Hastelloy G in 85% phosphoric acid at 160°C is only 0·5 mm/year, against some 42 mm/year for 316 stainless steel. Incoloy alloy 825 corrodes to the extent of 0·18 mm/year in boiling 70% phosphoric acid, as against a corrosion rate of 5·4 mm for 316L stainless steel.

A test was carried out with a mixture of phosphoric acid, sulphuric acid and hydrochloric acid, at elevated temperatures. All forms of stainless steel were badly attacked by this mix, but Hastelloy alloy C was found to be resistant. Hastelloy alloys B and C, as well as Inconel and Monel were also found to be extremely resistant to fumes containing such potent materials as hydrogen fluoride, silicon tetrafluoride, hydrofluosilicic acid as well as mists of sulphuric and phosphoric acids.

MARINE ENVIRONMENTS

When no currents are flowing, copper, 90/10 and 70/30 copper–nickel alloys are not susceptible to localised attack when immersed in sea-water, whereas under similar conditions, nickel and Monel metals are corroded. The picture, however, changes when there is a flow of water past the metal surface. Copper is attacked by sea-water as soon as the flow-rate past its surface exceeds 0·6 m/sec while 90/10 and 70/30 copper–nickel alloys are attacked at flow-rates above 4·5 m/sec, being resistant below this flow-rate. On the other hand, nickel and the Monel metals, which are attacked by stagnant sea-water, are resistant to impingement attack at flow-rates above 4·5 m/sec and retain this resistance right up to flow-rates of 42 m/sec.

GALVANIC ACTIONS

There is a difference of 0·08 V between the 90/10 copper–nickel alloy and the 70/30 copper–nickel alloy in sea-water, with the former being the more anodic. Consequently there is a slight galvanic effect when the two are coupled together. This does not matter when the surface area of the 90/10 alloy is large in comparison with that of the 70/30 alloy, or even of the same order of magnitude, but affects corrosion when the opposite is the case and the surface area of the 90/10 alloy is small. In particular one should never use a 90/10 welding electrode for joining 70/30 piping together. Heat exchanger tubes used with sea-water are often made of 70/30 alloy, clad with 90/10 alloy. Monel alloy 400 is cathodic by some 0·9 V to 70/30 alloy and by 0·17 V to 90/10 alloy. Consequently, sacrificial zinc anodes should be used whenever it is necessary to couple Monel 400 and one of the copper–nickel alloys together.

Chlorination of sea-water, which is sometimes carried out to prevent fouling, has no effect upon the corrosion rate of the copper–nickel alloys. Corrosion is, however, enhanced when the sea-water is polluted with hydrogen sulphide or when scaling takes place.

CHROMIUM–NICKEL ALLOYS TO OVERCOME FUEL ASH CORROSION

High-temperature corrosion in furnaces has become a growing problem for the following three reasons:

1 Furnace temperatures have become higher in recent years.
2 More efficient oil refining leaves less residual oil, which is therefore proportionally richer in vanadium, sodium and sulphur compounds.
3 There are financial advantages in using residual fuel oils rather than more refined products.

Corrosion proceeds due to the formation of a slag of vanadium pentoxide and sodium sulphate, which fluxes the protective oxide film that forms on heat-resisting alloys, so that oxidation and sulphidation can take place at a very rapid rate. High-temperature corrosion usually becomes serious between 600 and 750°C.

Two corrosion-resistant alloys are used: the 50/50 type consisting of 50% chromium and 50% nickel and the 60/40 type which consists of 60% chromium and 40% nickel. These alloys have been found to be very much superior to stainless steels and many other alloys for high temperature purposes. The 50/50 chromium–nickel alloy is easier to work, but at really high temperatures and aggressive conditions, the 60/40 alloy is superior. In a field trial it was found that a 60/40 chromium–nickel component was still useable after 17,000 hr under furnace conditions where maximum temperatures had gone up to 850°C and where fuels with a vanadium content of up to 0·4% and a sulphur content of up to 2·75% were burned. Stainless steel failed completely after 8,500 hr under such conditions.

Literature sources and suggested further reading

1 LA QUE, F. L., and COPSON, H. R., *Corrosion resistance of metals and alloys*, Reinhold, New York (1963)
2 *The corrosion resistance of stainless steel*, BISRA, London (1968)
3 BOOKER, C. J. L., *Corrosion of metals and alloys*, translated from the Russian, National Lending Library for Science and Technology (1967)
4 'Corrosion chart', *Chemical Processing* (December 1969)
5 LEWIS, H., PENRICE, P. J., STAPLEY, A. J., and TOWERS, J. A., 'Nickel-chromium alloys with 30 to 60% chromium in relation to their resistance to corrosion by fuel ash deposits', *J. Inst. F.* (January 1966)
6 *Chromium–nickel alloys*, International Nickel Company Limited (1966)
7 *Corrosion resistance of austenitic chromium–nickel stainless steels in marine environments*, International Nickel Company Limited (1964)
8 *Corrosion resistance of nickel-containing alloys in phosphoric acid*, International Nickel Company Limited (1966)
9 *Corrosion resistance of the austenitic chromium–nickel stainless steels in chemical environments*, International Nickel Company Limited (1966)
10 *Corrosion resistance of the austenitic chromium–nickel stainless steels in high-temperature environments*, International Nickel Company Limited (1966)
11 KIRKALDY, J. S., and WARD, R. G., *Aspects of modern ferrous metallurgy*, Blackie, Edinburgh (1964)

Chapter 7   Metal Coatings

Numerous metallic coatings are applied to base metals. These can protect in two ways:

1 A coating metal of a cathodic character serves to keep out air and moisture from the surface of the metal it covers. Such a coating is only of value if it is kept totally undamaged, as heavy pitting takes place at positions where the coating is punctured.
2 A coating metal anodic to the base metal only corrodes slowly itself due to the existence of considerable polarisation effects. The best known of such coatings is that of zinc (galvanising). In this case the protective action is retained even if the coating is punctured.

7.1 Application of coatings

The coatings themselves are applied by the following methods:

1 Electroplating.
2 Hot dipping.
3 Metal spraying.
4 Condensation of metal vapour.
5 Metal cladding.
6 Cementation.

Naturally, not all six methods can be applied with all metals.

ELECTROPLATING

This is perhaps the most important of all the methods of applying metal coatings on base surfaces.

It is of vital importance that the article to be plated is cleaned extremely carefully, and that no extraneous matter such as grease or oxides remain. To improve the adhesion of the coating metals it also often pays to roughen the surface of the base metal by sand-blasting and other techniques. After having been cleaned, the article to be electroplated is made the cathode of an electrolytic cell, while the plating metal is used in the form of rods or slabs as the anode.

The most crucial factor of electroplating is the nature of the electrolyte used. It is necessary that the metal be deposited in the form of a firm and dense coating, and not as a spongy deposit. Although the current efficiency of complex salts such as double cyanides, etc., is less than that of simple salts, it has been found by experience that the deposits formed are much better. Often numerous other agents such as colloidal additives are added to give the best results. The nature of the best electrolyte to use has been established empirically by extensive research backed up by practical experience and varies from metal to metal. It is also frequently governed by the shape and nature of the base metal.

The following troubles can arise with electroplating:

1 Deposits become spongy. This is due to the formation of metal hydroxides and may be caused by the electrolyte being insufficiently acid.
2 Pinholes appear in the film. The reason for this is either the release of hydrogen at the cathode surface (solution too acid), or the presence of particles of dirt in the solution. Bridging of pores takes place whenever coatings exceed 0·03 mm in thickness.
3 Cracking of electro-deposits. The reason for this is usually the presence of trapped water molecules at grain boundaries. This phenomenon can be avoided by slowing down the electro-deposition process, or alternatively using techniques such as ultrasonics to relieve internal stresses.

Electro-deposition produces purer and more even coatings than other techniques, and enables better control to be achieved. Wire and strip plating can be carried out in long plating baths, while small parts such as screws, nuts and bolts are plated in rotating perforated barrels.

HOT DIPPING

Hot dipping only works if the base metal forms an alloy with the coating metal. This is the case with such common hot-dipping processes as the coating of iron by aluminium, tin or zinc. Often it may be necessary to add a third metal to the liquid mix, to produce a ternary alloy. Lead does not form an alloy with steel, yet it is possible to coat steel with a lead film, if tin is also present.

Whenever base metals are coated in this way, there are always two distinct layers on the surface, namely the alloy layer and the layer of reasonably pure coating metal on top. The thickness of each is often governed by the period of immersion of the base metal in the coating-metal bath. Excessive immersion periods cause brittle coatings.

The advantages of hot dipping are that the plant used is relatively simple and it is easier to obtain good results than with, say, electrolysis. On the other hand, hot dipping tends to produce rather thick coatings and it is not too easy to control uniformity with irregularly shaped objects. Also hot dipping is only generally applicable to metals which have a low melting-point. At the time of writing, zinc, tin, aluminium and lead are the only metals where hot-dipping is of reasonable commercial importance.

METAL SPRAYING

This technique has expanded considerably in recent years, and is widely applied to the protection of large structures against corrosion. Aluminium and zinc are the most commonly applied by this technique, using an oxy-acetylene spray gun and the metal in the form of powder or wire. Exposed structural steel work is often protected by thicknesses of about 0·10 mm of aluminium which is applied in this way, while surface protection of bridges is commonly effected by zinc coatings between 0·05 and 0·08 mm thick, again applied by a flame-spray gun.

The newest technique of spraying not only metals, but also refractory oxides, carbides and nitrides on to the base metals is by the use of the so-called plasma gun. This uses nitrogen (or, when metals that react with nitrogen are used, argon), which is partially ionised in an electric arc. The gas is shot out through an orifice and achieves a temperature up to 8,000°C. Virtually all metals and refractory oxides can be sprayed on to base metal surfaces with a plasma gun, which causes the molten particles to adhere, because of the high impact velocity of 250 m/sec with which they strike the base metal.

The guns used for metal spraying usually weigh about 1·5–2 kg and it is possible for a skilled operator to cover 40–50 m²/hr with the appropriate metal film.

During the spraying process the globules of molten metal flatten out against the surface and there is also some flow into pores and irregularities, which locks the coating into place. Unlike hot dipping, there is usually little alloying between the base metal and the coating. Coatings are frequently somewhat oxidised and may be porous, unless thicknesses are substantially higher than would be the case with coatings applied by other techniques. The existence of pores in the coating is not particularly damaging in the case of sprayed zinc, because zinc is anodic to iron and tends to protect it. Where cathodic metal coatings such as tin and lead are concerned, the pores can be very harmful, and subsidiary processes are often used to eliminate them. Hammering, shot 'peening' and wire brushing of spray-applied tin and lead coatings have been found to improve the resistance of objects to chemical attack, because such treatment reduces porosity. Sprayed aluminium and zinc coatings are best protected additionally by painting. It is suggested that low-viscosity paints should be used to enable the pore structure of the spray applied metals to be easily penetrated.

CONDENSATION OF METAL VAPOUR

Three distinct methods are used for the attachment of metal vapours to base metals:

Metal evaporation
The vapours of metals can be condensed on the surface of base metal. This process is mainly carried out in an inert atmosphere for metals with a low melting-point such as aluminium, lead, tin and zinc. When more modern processes, with very high temperatures and a vacuum as low as a few millinewtons per square metre, are used, beryllium, cobalt, copper, gold, manganese, nickel and silver may all be deposited on base metals. It is essential to keep the temperature of the deposited metal within fairly narrow limits.

Cathode 'sputtering'

By using 500–2,000 V between two electrodes and maintaining a pressure of around 8 mN/m², a metal sublimate with a fine grained structure is formed. This can be deposited very evenly on base metals. Virtually pore-free films can be obtained whenever the thickness of the layer applied exceeds about 0·001 mm.

Thermal decomposition of metallic compounds

A typical example of this is the method of vapour nickel plating by heating nickel carbonyl. An object is heated to 180°C and a mixture of nickel carbonyl vapour and hydrogen is passed over it. Carbonyls of chromium, molybdenum and tungsten are also used in a roughly similar way. It is possible to attach uniform coatings, 0·05 mm or thicker, on base surfaces of steel, alloy steel or copper using this technique. Aluminium coatings are usually deposited from triisobutyl aluminium; tungsten coatings can also be obtained from tungsten hexachloride.

METAL CLADDING

Metal cladding is carried out by sandwiching the base metal between two layers of coating metal and then hot-rolling the composite, to produce firm adhesion. For example, an aluminium alloy such as Dural is cast into a steel mould which is lined internally with sheets of pure aluminium. By passing the ingot through a rolling mill, Dural sheeting coated on both surfaces by a film of pure aluminium is produced. This material has the excellent strength characteristics of Dural and also the superior corrosion resistance of pure aluminium. It is widely used in industry as 'clad sheet'. Similar methods are used for the production of steel plate clad on both sides by nickel, Monel metal, Iconel or stainless steel. The most widely used stainless-steel clad vessels have 6 mm thick linings of austenitic stainless steel on the inside.

CEMENTATION

Cementation is the process by which the base metal is heated while in contact with a powder of the coating metal to a temperature somewhat below the melting-point of either. A typical example of this is 'sherardising', by which process small steel objects are embedded in a rotating drum filled with zinc dust and heated to just below the melting-point of zinc. Apart from this, which is the most common example of cementation, the technique is also employed for coating iron or steel with the following metals: beryllium, boron, chromium, cobalt, manganese, molybdenum, silicon, tantalum, titanium, tungsten, uranium, vanadium and zirconium. All these coatings are applied at temperatures between about 800 and 1,400°C.

The mechanism by which this process takes place is as follows. At the temperature used the coating metal has an appreciable vapour pressure and, in addition, there is intimate contact between the solid phases. All this produces a diffusion of metallic atoms from the coating metal into the surface of the base metal to produce a surface alloy. It is essential that the surface be free from oxidation products as this tends to retard the cementation process. The coating produced is an alloy layer between the coating metal and the base metal and is therefore comparable to the inner or alloy layer of hot-dipped coatings for the same metals. Unlike other methods of coating metals there is no gradual change in composition from the base metal upwards, and there is no top layer of pure coating metal. The alloy layers are completely homogenous in each layer and, as a rule, cementation coatings are harder than pure metal coatings. Cementation techniques are also used, for example, for hardening of steel by the 'case hardening' and 'nitriding' methods. However, such techniques are mainly used to improve the strength and hardness of steel, rather than its corrosion resistance, and are outside the scope of this book. The main anticorrosion coatings applied by cementation are aluminium (calorising), chromium (chromising) and zinc (sherardising).

PRACTICAL METHODS

Cementation is best used for small components. These are packed into a completely gas-tight rotating drum, filled with the metal powder. Heating is usually by means of electric coils in the drum body. Calorising has to be carried out in a hydrogen atmosphere. Sherardising uses a mixture of 5% zinc powder and 95% alumina to prevent the zinc from caking. Although cementation-deposited coatings are not as corrosion-resistant as coatings applied by other methods, they have the advantage that the contours of machined surfaces remain intact. Cementation is the ideal way of treating nuts, bolts, screws and other similar small objects.

7.2 Various methods used for coating base metals

ALUMINIUM COATING

An aluminium coating does not offer any electrochemical protection to steel except in chloride solutions. Its value lies in the relative intertness of the coating itself, caused by the stability and coherence of the surface oxide film. The main methods of application are metal spraying, hot dipping, cementation and cladding. Aluminium can also be applied by vapour coating and by electroplating from a non-aqueous solvent. The latter method is still in the development stage.

The most common method of applying an aluminium coating is by hot dipping. The main iron–aluminium compound formed is $FeAl_3$ which is rather brittle and should be as thin as possible. To reduce the thickness of the $FeAl_3$ layer, about 2·5% of silicon is added to the molten aluminium. The steel has to be carefully surface treated before being coated. It is first heated to 430°C and then passed through a reducing atmosphere of hydrogen or cracked ammonia at 700°C, before entering the aluminium bath. It is also possible to use fluxes containing potassium fluozirconate or potassium fluotitanate. These are applied before passing the steel objects through the reducing atmosphere. The average thickness of aluminium coatings produced by these means is 0·06 mm.

Calorising is a cementation technique, in which articles are heated in a hydrogen atmosphere in contact with a mixture of aluminium, aluminium oxide and about 3% aluminium chloride. Iron and steel objects are heated to about 900°C and copper objects to about 750°C. After this stage has been completed, the objects are heated for about 24 hr at 950°C to produce a diffusion depth of between 0·5–1 mm and an external aluminium content of about 25%. The second treatment is necessary as the surface layer is somewhat porous after the initial calorising process and therefore gives poor protection. Calorised metal must not be severely deformed, otherwise the coating cracks; a 5% elongation is about the limit.

The vapour decomposition process consists in exposing steel objects to dry aluminium chloride vapour in a reducing atmosphere at about 1,000°C. The following reaction takes place:

$$2AlCl_3 + 3Fe \rightarrow 3FeCl_2 + 2Al$$

The coating obtained is somewhat similar to that produced by the calorising process.

Cladding of aluminium is widely used, both for steel base metal and for alloys. Bonding of aluminium to steel is effected by rolling at 540°C. Aluminium alloy base metal coated on both sides by pure aluminium is called 'Alclad' and is used in the construction of aeroplane fuselages.

A pilot plant constructed recently applies aluminium electrolytically from from aluminium chloride dissolved in a mixture of formamide, nitrobenzene and benzoyl chloride. The bath is operated at 50°C with a current density of 30 A/dm^2 and produces a coating of aluminium 0·07 mm thick. After immersion for one second in 5% HCl solution, the surface, originally brown in appearance, becomes bright.

Aluminium coatings give good resistance to atmospheric pollution, particularly where sulphur dioxide and hydrogen sulphide are present. They are unsuitable for the protection of underground structures. Halogen salts also attack aluminium coatings and therefore aluminium coatings are not recommended for use in contact with sea-water and other chloride solutions.

ANTIMONY COATINGS

Antimony is mainly deposited electrolytically from antimony fluoride solution containing sodium citrate, gluconic acid and hydrogen peroxide. Coatings about 0·02 mm thick are usual, and resist hydrochloric, sulphuric and hydrofluoric acids as well as salt spray fog. They are not recommended against nitric acid and other oxidising acids.

CADMIUM COATINGS

Cadmium resembles zinc in being anodic to iron or steel, but its protective power in this respect is not as good as that of zinc. Its advantages over zinc are that it is less liable to form white corrosion products and soldering with non-corrosive fluxes is possible. Cadmium is superior to zinc in resisting very humid internal atmospheres, but is worse for external use. In addition, cadmium plating is more expensive than zinc plating.

Cadmium plating is carried out commercially in cyanide baths containing a mixture of cadmium oxide and sodium cyanide to produce $Na_2Cd(CN)_4$. A typical formulation of a bath suitable for cadmium plating contains the following quantities per litre of solution:

Cadmium oxide	24 g
Sodium cyanide	75 g
Sodium hydroxide	15 g
Triethanolamine	20 ml
Nickel oxide	0·2 g

Plating is carried out at a current density of 5 A/dm². Where objection to the poisonous nature of cyanide baths makes these undesirable, an alternative plating solution that contains the following materials per litre of solution may be used:

Hydrated cadmium sulphate	50 g
Sulphuric acid	50 g
Gelatine	10 g
Sulphonated naphthalene	5 g

The current density to use in this case is 1·8 A/dm².

CHROMIUM COATINGS

Electrolysis

The most common way of applying chromium coatings is by electrolysis, although cementation (chromising) is also used.

Chromium plating is usually employed simply as a very thin veneer on top of nickel-plated surfaces.

Hard chromium is also applied directly on to aluminium, titanium or hard steel, but the purpose in this case is not protection against corrosion, but to increase the surface hardness.

The electrolyte used for chromium plating contains the following materials per litre of solution.

Chromic acid, H_2CrO_4	300 g
Sulphuric acid	2·6 g
Sodium fluoride	0·4 g
Sodium fluosilicate	0·2 g

Such baths can be operated at up to 90°C and at current densities of up to 30 A/dm². Lead anodes are usually employed, the chromium used up being replaced in the form of chromic oxide, CrO_3.

If the bath temperature is too low or the current density too high, or both, the deposit tends to be dull in appearance. Similarly if too high a temperature is used, then the coating is milky. A good, smooth and shiny chromium coating that requires no buffing is obtained only within a narrow temperature–current density band (see Fig. 7.1). If the layer of chromium is more than 0·003 mm thick it generally possesses micro-cracks, whereas if it is less than 0·3 μm thick, it tends to contain pores. Modern duplex chromium coatings are thicker layers of chromium, which are micro-cracked, deposited on a base layer of crack-free chromium, which in its turn is laid on to a nickel-plated base.

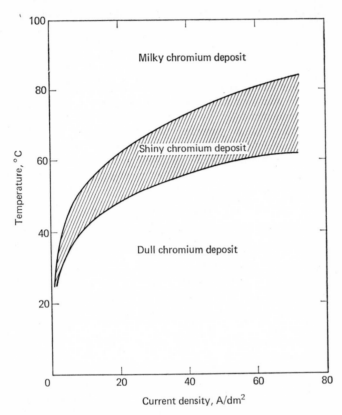

Fig. 7.1. Nature of chromium surface obtained under differing electrolysis conditions

The current efficiency of chromium plating baths is about 17% but this can be raised to 40% if hydrofluoric acid is used instead of sulphuric acid. The throwing power of chromium plating baths is low and trouble can occasionally be experienced of recessed areas being inadequately plated, or that the chromium deposit is milky.

Chromising
This cementation method uses 55% chromium powder and 45% Al_2O_3 at a temperature up to 1,350°C for 4 hr to provide a chromium alloy coating on small objects. The process is carried out in a hydrogen atmosphere, using a revolving electric furnace.

Chromising can also be done at a lower temperature (1,000°C) using a mix of chromium, alumina and ammonium chloride. The penetration depth achieved is about 0·1 mm after 4 hr; the surface contains up to 50% chromium after such treatment.

Chromium surfaces are coated immediately by a minutely thin oxide layer, which repairs itself when damaged. For decorative chromium plating, the thickness of the chromium layer on top of the nickel- and copper-plated base metal is usually about 0·5 μm. For car bumpers, etc., duplex coating which varies in thickness between 1 and 5 μm is usually employed. This has far better corrosion resistance properties, particularly in industrial atmospheres. Heavily chromium-plated steel is fairly resistant to steam, organic materials and flue gases. It is not recommended for use with reducing salt solutions or with reducing acids such as hydrochloric or sulphuric. Chromised steel is equivalent in its corrosion resistance properties to high chromium stainless steel, provided the surface layer is not destroyed. Chromising is found to be advantageous for small objects where a combination of surface hardness and improved corrosion resistance is desired, but where the additional cost of making the objects in stainless steel is not warranted.

COPPER COATINGS

Copper plating of steel is restricted in practice to two functions:

1 As an undercoating for the application of nickel and nickel–chromium layers.
2 For the production of components such as ground rods and anchor rods which are to be buried directly into the soil.

Copper electroplating baths contain, per litre, either:

$CuSO_4.5H_2O$	200 g
H_2SO_4	80 g
Phenolsulphonic acid	1 g

or:

$Cu(BF_4)_2$	300 g
HBF_4	20 g
H_3BO_3	20 g

Current densities of up to 10 A/dm² can be used.

If particularly fine grain coats of copper are required, small quantities of thiourea, sulphonated phenols and similar materials are added to the electrolyte solution.

Alkaline copper cyanide baths are also used, particularly for purposes where good throwing power and fast plating rates are needed.

Copper can also be applied to steel by a cladding technique in which steel is heated to red heat and immersed in molten copper. Subsequently the treated steel is rolled at 950°C. Such clad steel is non-porous and unless the coating is punctured, the object has the same resistance to the soil as a billet of pure copper.

GOLD COATINGS

Although gold plate was once considered solely for the purpose of jewellery manufacture, it is now used quite widely for protection against chemical attack. Gold is very unreactive, and because of this it is absolutely essential that coatings be completely non-porous, as otherwise rapid corrosion of the base metal may be induced. The main method of coating objects with gold films is by electrolysis. The electrolyte consists normally of a mixture of gold cyanide and sodium cyanide with a potassium carbonate buffer. The baths are operated at 60°C with a current density of 0·3 A/dm². Alternative electrolyte baths are available for specialised uses, including one containing a mixture of auric chloride and ammonium citrate. Such a bath can be operated at 60°C and 4 A/dm².

Gold can also be deposited chemically by the use of auric chloride solutions together with reducing agents such as sodium hypophosphite, tartaric acid or citric acid, but applications are limited. Finally, gold deposits can be alloyed with other metals to improve the toughness of the surface layer or to produce different shades of gold. The addition of potassium antimony tartate to a plating bath gives deposits of the harder gold–antimony alloy, while the addition of copper, nickel or silver salts to the plating solution also permits the coprecipitation of these metals in the gold coating.

LEAD COATINGS

The main advantages of lead coatings are the following:

1 It is the only reasonably cheap material that resists sulphuric acid fumes and sulphur dioxide.
2 It forms a good base for the application of paints.
3 It enables the base metal to be bent or otherwise deformed without cracking the coating.

Lead is most usually applied by hot dipping, but commercial methods of applying lead electrolytically have also been developed.

Hot dipping

Steel is first of all cleaned carefully and then dipped into a flux of a saturated solution of zinc chloride containing 5% ammonium chloride. The object is then dipped into a bath of molten lead at 350°C and the process repeated to give a very firm coating of lead. Somewhat harder coatings are obtained by adding about 2–2·5% of tin, antimony or cadmium to the lead in the lead bath. Lead-coated sheets are also produced from galvanised steel sheeting but it is necessary in such a case that the lead bath contains traces of zinc.

A common form of dipped-lead-coated steel is the so-called 'Terne' plate. This is made by dipping a continuous steel sheet in a zinc chloride flux and a bath containing a mixture of 80% lead and 20% tin. Terne plating is sold in terms of gauge numbers which were introduced by the ASTM (American Society for Testing and Materials). They are as follows:

L

Coating weight, g/m^2	Gauge
Below 80	10
100	16
135	18
170	20
Above 300	22

Terne plating is used for roofing, petrol tanks and for many purposes in the chemical industry where resistance to acid fumes is needed.

Electroplated lead coatings

The electrolyte bath most commonly used is one of lead fluoborate containing hydroquinone. Such a bath is operated with a current density of 100 A/dm^2 and can be used to deposit impervious lead coatings as thin as 3 μm.

Lead is also often applied on top of copper coatings, to obtain additional protection. A common formulation of such protection is a base layer of 0·4 μm of copper followed by a layer of lead between 10 and 20 μm thick.

Lead coatings resist the same chemicals as pure lead. Consequently, steel protected by a lead coating is widely used in the chemical industry, as well as for atmospheric exposure in industrial areas. Lead-coated steel is not recommended for marine environments, nor for use in permanent contact with the soil, particularly tidal marshes and loams, as pitting can be heavy.

NICKEL COATINGS

Nickel is one of the most important of the common coating materials because it has virtually the same strength, hardness and ductility as steel, yet is extremely resistant to corrosion. Nickel is most commonly applied electrolytically, but coatings can also be applied by chemical means or cladding techniques.

Electroplating

Plating baths have the following composition per litre:

Nickel sulphate	350 g
Nickel chloride	50 g
Boric acid	40 g

Such baths tend to produce rather dull surfaces of nickel. To get a bright nickel film on the surface of base metals it is necessary to add small quantities of such materials as naphthalene disulphonic acid, diphenyl sulphonate, benzene sulphonamide, etc., as well as zinc, cadmium or mercury salts. The current density used is about 8 A/dm^2. In the motor-car industry it is best to produce a semi-bright nickel deposit, which is well keyed to receive the chromium coating that follows, by some addition of cobalt salts. Electrolyte baths where some or all the nickel sulphate is replaced by nickel chloride enable a higher current density to be used and also often give better films though handling difficulties are very much increased due to the corrosive nature of such solutions. In all cases the anode consists of bars of pure nickel.

The control of the electrolytic process is somewhat critical. Steel surfaces must be very clean and the electrolyte must be filtered continuously during operation. Continuous agitation is also needed to prevent cathode polarisation and the precipitation of metallic ion impurities which are released into the solution, as the anode bars dissolve. Because different types of nickel deposit are required for different kinds of end product, the basic electrolytic process is modified by varying such factors as the pH of the solution, addition of traces of other metal ions, etc.

Apart from steel, many other base metals are nickel-plated commercially. Aluminium is mostly nickel-plated on to a zinc undercoat, or by an anodic treatment. Using an anodising process (page 169) a thick oxide film is formed, followed by dipping into a solution containing 100 g/litre ZnO and 450 g/litre NaOH. The aluminium oxide film is thus replaced by one of zinc, 0·6 μm thick, which serves as a base for the nickel film that follows.

Zinc-base die castings, magnesium, titanium, molybdenum, nobium and beryllium are all metals which are commonly electroplated to provide them with nickel coatings. Zinc-base castings are often given a thin copper undercoat, prior to nickel plating. As is usual, the final coating applied is a minutely thin one of shiny chromium. Zinc-base die castings that have been treated in this way are widely used in the motor-car and building industries.

Chemical deposition of nickel

Alternative methods are used where electroplating is found difficult to execute. Steel articles can be plated by immersing them into nickel sulphate or nickel chloride solutions at about 70°C and adjusting the pH to 3·5 by means of a buffer solution. After about 5 minutes' immersion, the object is washed in a dilute sodium carbonate solution and heated to 700°C. A disadvantage is that coating thicknesses are usually well below 1 μm. Such nickel films have been found useful as a basis for the application of ceramic coatings on steel.

Thicker deposits of nickel can be applied by using nickel chloride solutions in conjunction with sodium hypophosphate as a reducing agent, and an organic acid that serves as a buffer. Small quantities of suspended metals such as cobalt, palladium and aluminium catalyse the deposition reaction. Accelerators such as diaminobenzoic acid accelerate the rate at which nickel is deposited, while stabilisers such as glycine eliminate the effect of impurities. Nickel deposits up to 0·03 mm thick can be produced in this way. In order to harden the nickel films the object is usually heated in a furnace to about 750°C. It has been found that chemically deposited nickel is often superior to electrolytically deposited nickel in the resistance it gives to the base steel.

Nickel cladding

This technique is used in the chemical industry to produce containers able to withstand attack by alkalis, acids and a variety of salt solutions. Equipment of this type is used particularly widely in the food industry. Nickel cladding is applied by first grit blasting the steel to which the nickel is to be joined, and then welding the edges of the sheets together. Nickel and steel are then bonded by rolling at 1,200°C.

PALLADIUM, IRIDIUM AND PLATINUM COATINGS

Before coating objects with these metals it is customary to use a nickel undercoating, and a very thin layer of silver as a base. Palladium is deposited electrolytically using baths of palladate solution. It is possible to get non-porous deposits with layers only 0·5 μm thick. Platinum is commonly deposited by electrolysis from a platinate electrolytic bath, and iridium is applied by vapour deposition, using an electron beam. Coatings up to 0·2 mm thick can be applied in this way. Palladium and platinum can also be applied upon base metals by means of a cladding technique.

RHODIUM COATINGS

Rhodium has excellent anticorrosion properties, being resistant to acids, including aqua regia, chlorine and highly polluted atmospheres. Rhodium surfaces are also extremely abrasion resistant. Rhodium is generally applied to copper or silver undercoats from an electrolytic bath containing a mixture of rhodium sulphate and free sulphuric acid, as well as some selenic acid to induce even deposition. Plating is carried out at 50°C at a current density not exceeding 10 A/dm².

SILVER COATINGS

Silver is applied to base metals entirely by the use of electrolysis from a cyanide bath. A common industrial method involving silver plating is in the manufacture of ball and roller bearings, which are given a minutely thin coating of nickel followed by thin coatings of silver. Normally silver is applied to steel over a thin coating of copper. This produces silver plate, used in the cutlery industry. It has been found impossible to produce pore-free silver coatings on metal when thicknesses of silver applied are less than 0·025 mm. If silver is deposited on a copper or nickel undercoating which is free from pores even when applied in much thinner layers than this, it is possible to use silver thickness as low as 2·5 μm.

When applying silver to a base metal, a fairly concentrated silver cyanide–sodium cyanide solution is used as the basic electrolyte, to which 1 g/litre sodium thiosulphate and 10 ml/litre of ammonia are added as brightening agents. Some potassium carbonate is also added to improve the nature of the deposit. Electroplating is carried out at 45°C with a current density of about 10 A/dm².

TANTALUM COATINGS

Tantalum, which is a metal that resists many corrosive agents such as chlorine and strong acids, is applied by cladding, metal spraying or by electrolytical methods. Cladding consists simply of welding very thin sheets of tantalum to mild steel sections, followed by rolling. Metal spraying is not widely used, being restricted to such purposes as the protection of shafts and other steel components that are exposed to acids. It is essential to apply at least 2 mm of tantalum to base metals for surfaces to become continuous and non-porous.

Electrolytic methods are (at the time of writing) still somewhat experimental in nature, but it has been found possible to produce reasonably non-porous coatings of tantalum about 20 μm thick. To avoid the embrittlement of tantalum surfaces by hydrogen, which takes place in hot acids, it is common to coat the tantalum surfaces with an extremely thin film of platinum, or even to attack tiny pieces of platinum. As little as 0·01 of the surface area covered by platinum has a marked protecting effect upon tantalum.

TIN COATINGS

Tin has enormous advantages as a coating metal and for this reason is one of the most widely used. It is tough and adheres extremely well to base metals, being able to withstand extensive deformation. It is fairly non-porous even when applied in the form of thin coatings and has excellent corrosion resistance to the atmosphere and many other corrosive agents. Finally it can be used in contact with foods of all types because its salts are non-poisonous and non-toxic. Tin is applied to base surfaces by hot dipping and electroplating. Metal spraying and chemical means are also used, but applications are very rare. The main disadvantage of tin plating is that tin is cathodic to iron, so that when tin plate is scratched very rapid corrosion takes place at the scratches.

Electroplating
During recent years there has been a considerable expansion in the field of electroplating as against hot dipping because it is possible to produce much thinner coherent and non-porous films in this way. It is estimated that for low-cost tin cans as used in the food industry, tin thicknesses can be reduced by up to 50%. Some 98% of tin cans used in the food industry are now plated electrolytically.

Modern electro-tinning plants operate on a continuous strip basis, steel sheeting travelling along at a speed of about 8 m/sec, each coil of steel sheeting being welded to the next one to give continuity of operation. The steel sheet first passes through a bath of dilute sulphuric acid to remove scale. It is then cold-rolled, immersed in a hot alkali solution, then annealed at 650°C and rolled once more. These preparative methods are the same as those used for sheet steel coated by the former conventional hot-dip methods. Timing is carried out in either alkaline or acid baths.
Alkaline tinning baths consist, per litre of:

Sodium stannate	90 g
Sodium acetate	15 g
Sodium hydroxide	8 g
Hydrogen peroxide	0·1 g

Baths are operated at 75°C with a current density of 2 A/dm². Alternatively one can use a solution containing, per litre, the following:

Potassium stannate	400 g
Potassium hydroxide	20 g
Hydrogen peroxide	0·1 g

and operate such a bath at 40 A/dm².

Acid tinning baths contain 70 g/litre of stannous sulphate together with varying quantities of such substances as phenolsulphonic acid, fluoboric acid and traces of gelatine, glue, other phenolic compounds, etc., which help to provide good adhesion of the tin to the base surfaces. Electrolysis with acid baths proceeds at 50°C with a current density of 25 A/dm².

The tin deposited on the surface of the steel has a very dull appearance and must be melted in order to become bright and shiny. This is commonly carried out by the use of high-frequency induction heating using about 20,000 V at 2,000 Hz. The melting induced in this way is only very temporary, as otherwise iron–tin alloy formation, which reduces the corrosion resistance of the tin plate, may take place.

Finally, the tin plating is given a passivation treatment, using either immersion in hot chromic acid or electrolytic anodising with a mixture of chromic and phosphoric acids. Tin platings on cans of food are usually extremely thin, being of the order of 0·5 μm. Additional corrosion protection in the form of an applied lacquer coating is always given to electrotinplate used for the canning of food and drink.

Hot-dipped tin coatings

This method is now virtually obsolete for the production of tin cans for the food industry. It is used only for the production of heavier coatings on steel sheet for manufactured articles, intended to have a longer life than a food can which is discarded after one single use. The steel sheet is pretreated in the same way as for electro-tinning and passes through a layer of fused zinc chloride and ammonium chloride, which acts as a flux. It then goes through the tin bath (see Fig. 7.2) kept at 320°C and is covered on its surface by palm oil. The excess of tin adhering to the steel sheet

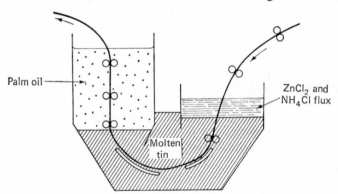

Fig. 7.2. Hot dipping for heavy tin coatings

is squeezed out by a set of rollers, which control the final thickness of the tin coating. Finally, the palm oil which still sticks to the surface of the tinplate, is removed by washing in hot alkali, followed by polishing with steel rollers covered with flannel material.

For tinning of irregularly shaped objects, such as castings, the objects are first immersed in a flux of molten sodium chloride and zinc chloride, followed by immersion in a tin bath covered by palm oil. Any excess of tin is removed by centrifuging.

Copper objects are tinned in the same way as iron and steel objects, and one of the most frequent applications of this is the tinning of copper wire used for electrical purposes. The tin layer protects the copper and the sulphur compounds contained in the rubber insulation layer from interacting with each other.

The nature of tin coatings

The extremely thin coatings of tin which are being applied commercially tend to be rather porous. It has been found that the porosity seems to vary inversely with the square of the coating thickness, amounting to about 600 pores/dm² with a coating thickness of 3 μm. Drastic reduction in the porosity of tin coatings can be provided by combining hot dipping with electrolysis. This is often carried out when shaped tinned objects are manufactured. These are made from hot-dipped tin plate and after manufacture they are re-tinned by electrolysis in an alkaline bath. This counteracts the severe porosity induced by bending the tin plate.

Tin resists atmospheric corrosion well because it becomes coated with a film of stannic oxide, which varies in thickness between 1·5 and 2 nm. This protects the rest of the tin against attack. The resistance of tin to seawater is also quite good.

When tin is used for the canning of fruits, which contain certain oxidants, it is possible for internal corrosion to proceed, often accompanied by swelling of the cans. The addition of gelatine to the contents of such cans tends to inhibit such effects. Many other foods such as meat, fish, etc., may also attack the tin coatings over long periods of time. Modern canning practice avoids such harmful effects by the use of vinyl, phenolic or wax coatings on the tin plating inside the cans.

Chemical deposition of tin

Very thin coatings of tin can be chemically deposited on to base metals. A typical example is the tinning applied on copper objects when these are immersed together with pieces of tin wire in a solution of sodium chloride and potassium hydrogen tartrate at 90°C for 4 hr.

Steel objects can be tinned chemically by immersion in a hot solution of stannous chloride, hydrochloric acid and sodium sulphate together with strips of zinc.

TITANIUM COATINGS

Titanium coatings are somewhat uncommon and are normally applied only electrolytically. A mix of 90% potassium titanium fluoride and 10% sodium

fluoride is melted and kept at 850°C. Using a current density of 25 A/dm²
it is possible to apply smooth, shiny and pore-free coatings of titanium
up to 0·05 mm in thickness.

Titanium coatings are used for the protection of steel against corrosive
chemicals and also against sea-water and salt sprays. The chemical properties of titanium coatings are similar to those of the metal (see Chapter 5).

TUNGSTEN COATINGS

Tungsten coatings are usually applied electrolytically from a fused electrolyte consisting of a mixture of sodium tungstate, lithium tungstate and
tungstic acid. Baths are kept at 1,000°C and the current density maintained
is about 75 A/dm². The deposits applied are usually up to 0·5 mm in
thickness. Tungsten can also be applied by thermal decomposition from
chemical compounds such as tungsten bromide and tungsten hexafluoride,
or by means of a cementation method. Iron and steel objects as well as
copper alloys are coated with an extremely thin tungsten surface layer by
heating to about 1,000°C in the presence of ammonium chloride vapour,
using a tungsten–iron alloy as the metal source.

ZINC COATINGS

Nearly half the zinc of commerce is used for metal coating purposes, or
galvanising as it is called. Zinc coatings are mainly applied by hot dipping
or electrolysis, but spraying, vapour deposition and cementation are also
quite widely used.

Zinc plating is probably the most useful of all methods of protecting
steel surfaces against corrosion, because zinc itself is fairly immune to
atmospheric attack due to strong polarisation effects, while it always remains
anodic to iron. This means that, unlike tin or most other metal coatings,
zinc protects the underlying steel even if the zinc coating should be scratched
or otherwise defective. The disadvantages of zinc coatings are the darkening

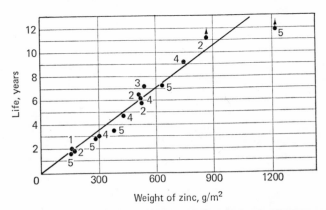

Fig. 7.3. Expected service life of various types of zinc coating in an industrial
atmosphere *(By courtesy of the Zinc Development Association)*

of the surface with the passage of time, producing a rather ugly rough surface, and the fact that it is readily attacked by acids and alkalis. Zinc salts are rather toxic, and zinc-coated cans must never be used in conjunction with foodstuffs. The limit of zinc salts permitted in the USA in drinking water is given as 5 ppm by weight of Zn^{2+}.

ELECTROLYTIC PROCESSES FOR COATING BASE METALS WITH ZINC

Electro-deposition is mostly used for irregularly shaped articles that are difficult to finish by the more usual hot-dip galvanising process. It is also used for applying very thin coatings of zinc to steel strip that are to be phosphated and painted afterwards. Electroplating can produce zinc coatings as thin as 1 μm.

The base metal has to be carefully cleaned, particularly when acidic plating baths are used. Electrolysis proceeds from either acidic plating baths or alkaline cyanide ones. Acidic baths produce rather coarse deposits and tend to coat complex objects unevenly. They can be operated at a high current density and are cheap. A typical formulation, per litre for an acidic electrolytic bath is:

Hydrated zinc sulphate	250 g
Ammonium chloride	15 g
Hydrated aluminium sulphate	30 g
Glycerol, dextrin, glucose, phenols, etc.	Traces

Such baths are operated at room temperature with a current density between 2 and 10 A/dm², depending upon the degree of agitation of the solution.

Numerous other acidic plating bath formulations are also commonly used. Alkaline zinc baths are highly poisonous, and for this reason there are stricter regulations for handling them. A typical formulation, per litre, is the following:

Zinc cyanide	60 g
Sodium cyanide	25 g
Sodium hydroxide	50 g

Such baths are operated at about 50°C with a maximum current density of 2 A/dm², and give a fine and even deposit on irregularly shaped articles. Brightening agents such as molybdic acid, tellurium salts or organic materials such as terpinol are often added to the electrolyte. Anodes used with both acidic and alkaline plating baths are of zinc. These are removed from the acidic baths when not in use, because they are attacked by the plating solution. Zinc coatings applied by the acid process are always rather coarse grained and have a grey finish. Coatings applied using an alkaline cyanide bath are usually much brighter, though both tend to darken as they age.

Hot dipping

Hot dipping of irregularly shaped objects is usually done by hand, or by semi-mechanical means. Continuous galvanising lines are used when steel strip, wire or wire netting is to be galvanised.

The steel object is first pickled in dilute sulphuric and hydrochloric acids.

This is quite a complicated process and requires a number of pickling baths, one placed behind the other. Organic inhibitors that retard the evolution of hydrogen by cathodic polarisation are also needed. Pickling can also be carried out by immersing the steel objects in a sulphuric acid bath that contains a small quantity of stannous sulphate. Tin is then deposited on the cleaned areas and stops the iron from going into solution. If the object to be descaled is made the anode in a bath of dilute sulphuric acid, cleaning of the steel surfaces takes place rapidly and efficiently. After pickling, the object is carefully rinsed in water prior to coating the surfaces with flux. The flux used is usually a mixture of zinc chloride and ammonium chloride which can be employed either as a hot solution prior to the zinc bath, or as a molten layer floating on top of the zinc. The purpose of the flux is to release a trace of hydrochloric acid on contact with the molten zinc, which effectively dissolves any oxide films that may have formed on the surface of the steel. The zinc bath is kept at about 450°C and steel objects are normally in contact with the molten zinc for about 5 min. Any excess of zinc is removed by a variety of methods. In the case of wire, asbestos pads are used to wipe off the excess; for steel sheet or strip the thickness of the zinc coating is adjusted by the use of rollers as with tin plating. With small objects the excess of zinc is thrown off by the use of a centrifuge, and with larger objects simple drainage combined with vibration is used.

Various additional metals are often added to the zinc baths in order to modify the properties of the coatings applied. Aluminium serves to produce a more even coating on irregularly shaped objects, whereas tin improves the appearance and degree of adhesion.

Sherardising

This process was invented by Sherard Cowper-Coles in 1900, and is used for the surface protection of small objects by coating them with a minutely thin layer of zinc-rich alloy. The process can be used for both steel and non-ferrous metal objects. The objects are packed into a drum together with zinc dust and alumina, and the drum is then rotated. The temperature inside is kept at about 360°C. Such a process provides a film of 25 μm after about 2–3 hr. The coating consists of a number of iron/zinc compounds and of solid solutions of these compounds with zinc.

METAL ALLOY COATINGS

Apart from coating base metals with *pure metals*, a number of *alloys* are also deposited by various methods.

Alloys of tin with cadmium, copper, lead, nickel and zinc are widely used instead of tin plating. These alloys are deposited electrolytically and possess advantages such as increased hardness, ease of soldering, etc., over pure tin coatings.

Brass coatings are applied by the use of cyanide solutions containing a mixture of copper and zinc cyanide, together with free sodium cyanide and sodium hydroxide. The anodes normally used contain 75% Cu and 25% Zn.

The chemical bath used for the deposition of *bronze* normally contains a mixture of potassium stannate and copper cyanide with an excess of potassium hydroxide and potassium cyanide. Several other additives are also necessary to give a satisfactory result.

Tin–lead alloys are used as a substitute for Terne plate and it has been found that a 6% tin coating produces the best anti-corrosion results.

Nickel–tin and *Nickel–cobalt alloys* are becoming of considerable commercial importance for metal protection. A 65% Sn, 35% Ni coating is being used instead of chromium plating on car components and for watches. Nickel–cobalt produces very hard and resilient plating. Both alloys are deposited electrolytically from plating solutions containing mixtures of the two cations.

Literature sources and suggested further reading

1 *Preventing the corrosion of steel in supply waters*, BISRA, London (1969)

2 BAKHALOV, G. T., and TURKOVSKAYA, A. V., *Corrosion and protection of metals*, Pergamon Press, Oxford (1965)

3 BURNS, R. M., and BRADLEY, W. W., *Protective coatings for metals*, Reinhold, New York (1967)

4 FONTANA, M. G., and GREENE, N. D., *Corrosion engineering*, McGraw-Hill, New York (1967)

5 *Proceedings of the Third International Congress on Metallic Corrosion*, Sweys and Zeitlingen, Amsterdam (1970)

6 GREENWOOD, J. D., *Heavy chemical and electro-deposition*, Robert Draper, Teddington (1970)

7 *Proceedings of the 8th International Conference on Hot-dip Galvanising*, Ind. Newspapers Ltd., London (1967)

8 BONNER, P. E., and WATKINS, K. O., *The priming of sprayed aluminium and zinc coatings on steel*, BISRA, London (1969)

9 DREWETT, R., 'Diffusion coatings for the protection of iron and steel' *Anticorrosion* (April, June and August 1969)

10 *Zinc coatings*, Zinc Development Association, London (1969)

11 LA QUE, F. L., and COPSON, H. R., *Corrosion resistance of metals and alloys*, Reinhold, New York (1963)

Chapter 8   **Paints and Varnishes**

Paints and varnishes are the most widely used anti-corrosion coatings on the surfaces of metals. They serve several functions:

1 They constitute a continuous coating that keeps the surface of the metal out of contact with air and moisture and therefore restrict corrosion.
2 The pigments present in the paint often act as corrosion inhibitors by electrochemical and other means.

Paints are made by mixing the various components intimately with each other.

8.1 Main constituents of paints

The main constituents of paints are as follows.

VEHICLE OR MEDIUM

This is a clear varnish or gummy base which attaches itself to the surface of the substance to be coated and solidifies there. Some vehicles dry simply by the evaporation of the solvent, but most harden by some chemical action such as polymerisation.

SOLVENT

The purpose of the solvent is to provide the surface coating material in a form in which it can readily be applied by brushing, spraying or dipping. The solvent is designed to evaporate after the surface coating material has been applied. Solvents vary in nature from water to such organic materials as esters, hydrocarbons, terpenes and ketones.

PIGMENTS

There are numerous inorganic and organic pigments, all of which are in the nature of insoluble white or coloured materials, chosen for their opacity. Pigments vary in their chemical properties, however, and must be carefully selected from the point of view of inertness to various surfaces and atmospheres.

EXTENDERS

Extenders are substances which are added in order to increase the 'body' of paints and thus to improve the abrasion and other mechanical resistances. Extenders are also used in matt paints. They are normally non-self-colouring.

DRIERS

Driers are catalysts that are often added to paints and varnishes in order to increase the rate of setting and to improve the final strength.

DYES

When surface coatings are intended to be completely transparent but to give a coloured tint, dyes are used instead of pigments. Surface coatings which incorporate dyes are not, however, very common.

When an organic surface coating contains a vehicle and solvent only, it is termed a 'varnish'. Paints are varnishes, that have pigments and/or extenders dispersed.

8.2 Vehicles or media

Vehicles fall into two classes: non-convertible materials and convertible coatings. The non-convertible type is one which simply dries on the surface, after the solvent has evaporated off. The most common of these are: Shellac, Manila gum, bitumen, waxes, cellulose acetate, non-drying alkyd resins, cyclised and chlorinated rubber and vinyl and acrylic resins.

In the case of the *convertible* materials, once the substance has been deposited on the surface of the object in question, its nature changes. Either under the action of atmospheric oxygen, or by the action of catalysis or heat, the molecules join up to produce large and tough polymeric films. The simplest of such materials are the various drying oils, which are unsaturated fatty esters. Typical examples of such drying oils are: linseed oil, tung oil, oiticica oil (China wood oil), dehydrated castor oil and soya bean oil, as well as oils made by blending and heat treatment of the raw oils.

The second group comprises a whole series of natural and synthetic resins, which can either be used neat, or reacted together with a drying oil in order to produce a more adherent and flexible substance. The main resins used in the surface coating field are the following: rosin, coumarone resin, zinc rosinate, calcium rosinate, copal resin, maleic acid/drying oil compounds, alkyd resins and alkyd resin/styrene copolymers, urea formaldehyde, melamine formaldehyde phenol formaldehyde and its copolymers, epoxide esters and polyurethanes.

Most surface coatings use a blend of synthetic resins in order to obtain the best results for each specific case.

DRYING OILS

Practically all common vegetable oils are mixtures of the triglycerides of fatty acids. But only if these fatty acids have one or more double bonds within their structure can the oil dry by oxidation and consequent film formation. The quantity of oxygen used for this setting process is found to be about 30–40% of the weight of the oil and is very much accelerated by the presence of so called 'driers'. For example, if 0·05% of cobalt in the form of cobalt naphthenate is added to linseed oil, it sets in 20% of the time needed for the drying of the neat oil. The rate of drying is also governed by the prevailing temperature.

A drying oil that is somewhat thicker than the raw oils is produced by a technique called thermal polymerisation. If linseed oil is heated for a number of hours at about 285°C it thickens appreciably and is then called 'stand oil'. Blown oils are made by blowing air through the raw oils at a temperature of about 120°C. This also thickens the oil, but the blown oils are not as durable and will tend to yellow more quickly than equivalent stand oils.

OLEORESINOUS VARNISHES

Clear film-forming liquids which dry by ordinary atmospheric oxidation are known as oleoresinous varnishes and form a hard and glossy film. All these varnishes contain drying oils, one or more resins, thinners and driers. The function of the oil is to give the varnish durability on exposure and flexibility. The resins used are hard and oil soluble, and are dispersed into the oil by heating. Their function is to improve the hardness and the gloss of the varnish. If a suitable resin is used, faster drying can be obtained as well.

The main thinners include white spirit, dipentene, coal tar naphtha, xylol and various petroleum spirits.

Small quantities of wood turpentine are also often added to produce a pleasant smell for the paint. The most common thinner of all is white spirit. Heavier petroleum distillates such as kerosene retard the rate of drying but improve the flow. More volatile solvents such as xylol and petroleum distillate are used where fast drying is essential. The quantity of solvent used depends upon the viscosity required. Brushing lacquers usually have a viscosity of 2 P at 25°C.

8.3 Driers

The most common driers employed are naphthenates, linoleates and rosinates of lead, cobalt (see Fig. 8.1), calcium and manganese. Lead driers induce polymerisation in the body of the film, whereas cobalt tends to induce primarily surface skinning. A wrong drier balance causes uneven drying of the paint film and may be the reason for surface wrinkling or crazing. Satisfactory drier ratios have been found to be: 0·05% Co, 0·5% Pb and 0·25% Ca for long alkyd varnishes and 0·05% Co, 0·025% Mn and

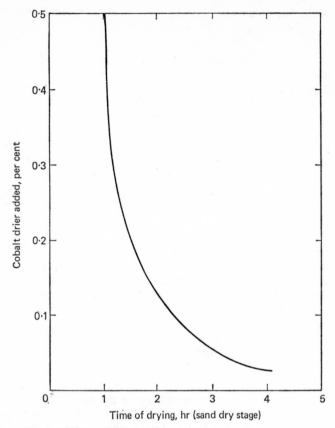

Fig. 8.1. Effect of drier addition on the drying time of a paint

0·25% Ca for medium length linseed oil/tung oil/alkyd clear varnishes. Each type of varnish requires its own specific drier balance, which is usually arrived at by methods of trial and error.

8.4 Classification of varnishes

An important criterion of varnishes is the resin–oil ratio after the solvents have been driven off. The usual terms used are:

Resin to oil ratio = 1 : ½ to 1 : 2 = short oil varnish
Resin to oil ratio = 1 : 2 to 1 : 3 = medium oil varnish
Resin to oil ratio = 1 : 3 to 1 : 5 = long oil varnish

Short oil varnishes dry quickly to form a hard film with a high gloss, but do not have very good flexibility. They are mainly used for interior purposes. Long oil varnishes dry slowly, but have excellent flexibility and

weather well. Such varnishes behave well when exposed to external corrosive atmospheres.

Alkyd resins

Alkyd resins are made by heating together polyhydric alcohols such as
glycerol with dibasic acids such as phthalic anhydride at a temperature of
about 220°C. The reaction is stopped before complete cross-linking of the
molecules has taken place. Such alkyd resins are, however, never used on
their own but are modified by reaction with linseed oil. These linseed-oil-
modified alkyd resins are nowadays the most widely used of all in the paint
industry and constitute more than half of all resins used.

When light pigments are required in the paint, it is necessary to use
soya bean oil or sunflower oil instead of linseed oil or China wood oil, as the
latter tend to yellow with age.

Alkyd–styrene copolymer resins

The effect of reacting styrene with an alkyd resin is to produce a material
capable of being incorporated with a smaller quantity of oil and yet be
compatible with solvents. These copolymers are quicker drying, harder and
also impart a pale colour. They are found in industrial finishes where pale
colour and rapid drying either at room temperature or on heating are
needed. Disadvantages are a certain lack in flexibility and a tendency to
be readily attacked by hydrocarbon solvents.

Amino resins

Urea-formaldehyde resins and melamine-formaldehyde resins are used
almost entirely for application as industrial stoving finishes, being employed
for coating the outsides of refrigerator cabinets, radiators, metal building
components, etc. These resins are cured speedily, are hard and possess
good colour-retention.

Epoxide resins

Epoxide resins are rather expensive and are employed in conjunction with
phenolics and urea-formaldehyde plastics for the manufacture of stoving
finishes with good flexibility. If ethylenediamine is mixed with the epoxide
finish just before application, polymerisation takes place rapidly at room
temperature.

Polyurethanes

Polyurethanes are compounds of polyesters with a rather low acid value
(less than 10 mg KOH/g) with tolylenediisocyanate. The isocyanate group
is very active and reacts quickly with materials containing active hydrogen
atoms such as castor oil, epoxy resins and many other plastics. This reaction is dependent upon temperature and the nature of reacting group.
Polyurethane coatings dry and reach the fully cured stage at room temperature, with the exception of the isocyanate products which are compounds

of the isocyanates with materials such as phenol or acetoacetic ester. These cannot set at room temperature, but if they are heated above 145°C, they split up and free isocyanate is released. This then immediately reacts with the polyester resin to form a hard and fully cured material.

When formulating urethane lacquers it is necessary to choose solvents that do not contain free hydroxyl or similar groups as they can react with the isocyanates. The main solvents used are ketones, butyl and amyl acetates, with toluene, xylene and naphtha as diluents. Urethane surface coatings have poor flow properties so that is is necessary to use flow-out agents such as ethyl cellulose or cellulose acetobutyrate in the formulations. Pigments chosen in connection with urethane finishes should be dry, as water reacts with isocyanates.

Polyurethane finishes have excellent chemical resistance, particularly to oils, solvents and ozone. They can be used satisfactorily up to 150°C.

Silicones
Silicone resins are usually supplied as a 50–80% solution in aromatic solvents. They are thermosetting resins and need curing at about 230°C. After they have set, silicone finishes can be used at temperatures up to 300°C. Paints based on silicones are virtually unaffected by weathering, and are not attacked by most dilute acids and bases and many other chemicals. On the other hand, silicone finishes do not adhere too well, lack toughness and are not very solvent-resistant. Silicone paints are used for painting steel chimney stacks and for similar high-temperature purposes.

Other resins used
Apart from phenolic resins which are used in oleoresinous varnishes only, some use is made of various unsaturated polyester resins for true 'twin-pot' resins in which the resin and the initiator are sprayed simultaneously from twin jet spray guns, to set on contact. It is possible, by this technique to obtain films that are of the order of 0·2–0·5 mm in thickness. The reason why this can be done is that solvents are not necessary with this technique. Polyamide resins are sometimes used in conjunction with other resins, such as epoxy, to produce films that have good flexibility, adhesion and impact resistance.

8.5 Corrosion-resistant water paints

Latex water paints consists of high-molecular-weight polymers that have been polymerised in the emulsified state. The most common of these is either polyvinyl acetate or polyvinyl acetate–styrene–butadiene copolymer. The polymer is usually in the form of particles between 0·5 and 1 μm in diameter. If the particle size is larger than this, gloss is impaired, and there is a danger of sedimentation. In addition, a small particle size means that the emulsion has better water-resistance and gloss.

When polyvinyl acetate emulsions dry, the polymer tends to form pow-

M

dery particles unless properly plasticised with substances such as dibutyl phthalate or tricresyl phosphate. In addition, water paints often have thickeners added which are basically hydrophilic colloids, e.g. sodium carboxymethylcellulose, various water soluble salts of polyacrylic and methacrylic acid and methylcellulose. The thickeners increase the consistency of the emulsion paint and affect the flow.

Finally there are some polyacrylic and polystyrene base emulsion paints on the market, both of which produce durable water resistant films when dry.

8.6 Pigments

Pigments are solid materials, which are suspended in the form of small discrete particles in the medium, and which serve to give the paint:

1 Colour.
2 Opacity.
3 Increased abrasion resistance.
4 Certain chemical properties.
5 Improved body.

Pigments can either be natural products that have been finely ground during manufacture, or they can be synthetic materials precipitated from solutions. They must at all times be individually suspended in the paint medium. This is done by a colloidal grinding technique. The pigment particles in the unwetted state tend to stick together and therefore a considerable amount of work has to be done on them in order to disperse them in the medium (see Fig. 8.2). The main types of grinders used in the paint industry to achieve this dispersion are the ball mill, the single roller mill and the triple roller mill.

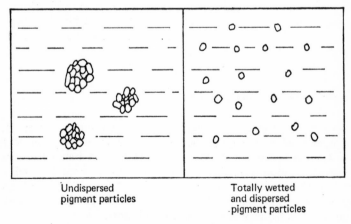

Undispersed
pigment particles

Totally wetted
and dispersed
pigment particles

Fig. 8.2. Grinding of a pigment into a medium

Pigments can be divided most conveniently into the following types:

1 Whites.
2 Extenders.
3 Inorganic coloured pigments.
4 Organic and lake pigments.
5 Metal powders.

WHITES

Titanium dioxide

This is the commonest of all white pigments and is manufactured in two modifications: anatase and rutile. Its great advantages as a pigment are its very high refractive index (2·52 for anatase and 2·76 for rutile), chemical inertness and lack of toxicity. The average particle size of TiO_2 pigments is about 0·25 μm. Rutile and anatase are different crystal forms of TiO_2 and have the following contrasting properties:

	Rutile	Anatase
Tinting strength	1,600	1,200
Density	4·2 kg/dm³	3·9 kg/dm³
Colour	Slightly cream	Pure white
Resistance to chalking	Very good	Poor

In consequence, titanium dioxide with a high anatase content is best for glossy paints which are used internally, while matt paints, especially those to be used externally should have a much higher rutile content.

Titanium dioxide is used in every kind of paint with the exception of very dark coloured ones. It is particularly useful for paints that are used in circumstances where contact with food is likely.

Zinc oxide and zinc sulphide pigments

Zinc oxide has a refractive index of 2·08 and zinc sulphide one of 2·37. The density of both pigments is around 5·6 kg/dm³.

Zinc oxide is brilliant white, but has a much poorer covering power than titanium dioxide. In addition it is a relatively reactive pigment and when used with oleoresin finishes is often the cause for cracking of the film. Its main advantage is its property of being able to absorb ultra-violet light. This property is extremely advantageous for external paints in which delicate organic lake colours are used. The zinc oxide aids both colour retention and stability of the medium film. Zinc oxide on its own is non-toxic but it is frequently used in the leaded form, i.e. containing quantities of $PbO.2PbSO_4$. Such pigments must, of course, never be used where likelihood of contact with foodstuffs occurs, or where children may come into contact with the paint film. Zinc oxide is somewhat fungicidal and is thus useful in retarding the formation of mildew on paint surfaces. It also retards putrefaction in glues, adhesives and water paints. The material must never be used with acidic paint media as it will cause such media to 'thicken' prematurely.

Lithopone
This is made by the following process:

$$ZnSO_4 + BaS \rightarrow ZnS + BaSO_4$$

to produce a pigment which contains about 30% zinc sulphide. The material is then roasted to produce a pigment with an average particle size of 0·5 μm.

Lithopone is a brilliant white pigment with excellent covering power. Its main use is in water paints and washable distempers for internal uses, but it is also widely used in all types of paints, rubbers, plastics and flooring compositions. Lithopone should not be used for external purposes because it is liable to chalk. This is then followed by weather erosion.

Antimony oxide
Antimony oxide, Sb_2O_5, has a refractive index of 2·0 and a density of 5·7 kg/dm³. Its main use in paints is to improve the brushing and levelling properties.

EXTENDERS

These are materials, which although they are usually white in colour, have a refractive index that is too low to contribute appreciably to the covering power of a paint. The reasons for the addition of extenders are the following:

1 They give body and improved brushing properties to a paint.
2 They increase the pigment content so as to produce a matt or semi-glossy surface suitable for undercoats.
3 Some of the extenders help to keep the pigment in suspension.

The main extenders used are as follows.

Paris white or CaCO₃
This is a pigment which, when dispersed in oil or cellulose media, gives a dirty colour to the resulting paint. It can, however, be used in water paints and distempers as a medium. As calcium carbonate is alkaline, it cannot be used in conjunction with acid pigments.

Barytes and blanc fixe
These are both chemically the same, namely barium sulphate $BaSO_4$. Barytes is the name given to the ground natural material, while blanc fixe is precipitated chemically. The latter has a finer crystalline shape. The disadvantage of both is the relatively high density of 4·6 kg/dm³, which tends to cause suspension difficulties.

Silica, slate powder and mica are all used as extenders in paints, where colour is not of primary importance. These materials are cheap and easy to incorporate.

Asbestine and talc are hydrated magnesium silicates with a theoretical formula of $H_2Mg_3 (SiO_3)_4$. The density of the material is about 2·8 kg/dm³ and the refractive index 1·59. Both materials are widely used as suspending agents in paint and for anticorrosive surface coatings.

COLOURED INORGANIC PIGMENTS

There are two main types of inorganic blue pigments, which are:

1 Prussian blue, made by reacting sodium ferrocyanide with iron salts. It is essentially $Fe_4[Fe(CN)_6]_3$ and is a strong pigment which is wholly unstable against alkalis. It must therefore never be used on setting concrete or plaster.
2 Ultramarine blue is a complex silicate which has a rather pinkish cast. It is stable towards alkalis but is attacked by acids. It is widely used in distempers and other water paints.

IRON OXIDE PIGMENTS

There is a large range of these pigments, which contain various oxides of iron together with additions of manganese dioxide. The colours are governed not only by the actual chemical composition but also by the state of subdivision, the crystalline forms, etc. The main iron oxide pigments used in the paint industry are red oxides of iron, yellow ochres, light brown Siennas and dark brown umbers.

All iron oxide pigments are very durable and have good colour retention. They are chemically inert. They are able to absorb ultra-violet light and thus tend to protect external paint films. Iron oxide is not a corrosion inhibitive pigment but improves the film characteristics of paints that include reactive pigments such as red lead or chromates.

Chromate pigments
A series of chromate pigments is used, of which the most important are the following:

Primrose chrome: 50% lead chromate, 50% lead sulphate
Lemon chrome: 75% lead chromate, 25% lead sulphate
Middle chrome: 100% lead chromate
Orange chrome: basic lead chromate
Scarlet chrome: lead chromate, lead sulphate and lead molybdate

All the lead chromates have the general disadvantages of lead pigments: poisonous nature and liability of blackening in the presence of hydrogen sulphide. Chrome pigments that do not suffer from these disadvantages are zinc chromate and barium chromate, which also have a strong yellow colour.

Black pigments
The great majority of black pigments used are carbon blacks, prepared by burning carbonaceous substances and allowing the flue gases to impinge

upon cold surfaces. Several varieties of carbon black are on the market, their names indicating the carbonaceous substances used in their manufacture, e.g. lamp black, vegetable black, bone black and furnace black.

A typical furnace black has the following properties:

Density	1.8 kg/dm^3
Average diameter of particles	$30–60$ μm
Volatile matter	$0.5–1\%$
Surface Area	$40–90$ m^2/g
pH	9.5

Carbon blacks have immense staining power but are hard to grind. Because they are so very light, they tend to float to the surface of paint films and in consequence paints that contain carbon black in addition to other pigments tend to darken on standing. For this reason it is difficult to carry out accurate colour matching when using carbon black. If accurate tints are needed, organic lakes are used instead.

Graphite is also used to some extent as a black pigment, but only where its protective action against corrosion is to be utilised. It has a poor tinting strength and a low oil absorption figure.

ORGANIC PIGMENTS

Organic dyes owe their colour to the fact that within the molecules there is a type of vibratory motion, which comes within the visible spectrum. Many of the common organic dyes dissolve readily in oil, resin or varnish and are used as such. Under such circumstances no opacity is contributed, and the result is simply a coloured transparent film.

Some directly produced dyes are quite insoluble in water or organic materials and can be used directly as pigments. Typical of such pigments is the well known blue copper phthalocyanine (see page 174), made by reacting phthalic anhydride, urea and cupric chloride.

Lakes

Lakes are made by reacting typical organic dyes, which bear an active chemical group, with a carrier material, which may either be basic or acidic. There are many hundreds of organic lakes manufactured with widely varying colours and properties. Details should be sought from specialist literature.

PIGMENTS USED SPECIALLY BECAUSE OF THEIR ANTICORROSION PROPERTIES

Inhibitive pigments are used in priming paints, i.e. in direct contact with the metal to be protected. In general, priming pigments are unsuitable for top coats. The following are the main pigments used for such purposes.

Red lead, PbO.PbO₂

This is made by roasting lead oxide at about 350°C, and the normal commercial grade contains about 97% Pb_3O_4. The particle size of the pigment is about 5 μm. Priming coats of red lead in the medium are only effective provided there is a high percentage of red lead in the paint. Red lead inhibits corrosion both by virtue of its alkalinity, which is caused by the presence of PbO, and by virtue of its oxidising nature. In addition, the presence of a lead salt causes the formation of a very dense paint coating on the surface of the steel, thereby reducing porosity.

White lead, $4PbCO_3.2Pb(OH)_2.PbO$ and calcium plumbate, $CaPbO_3$
White lead tends to be used as a top coat for paints required to protect external steel structures against corrosion. It is also used in priming coats, but is not as effective as red lead, as its protective power lies only in the alkaline salts which it forms with the medium. Calcium plumbate pigment is excellent for use with priming coats on steel. It has been found that this pigment is both cheaper and more effective, for priming paints used on galvanised steel, than red lead. Calcium plumbate inhibits the action on cathodic areas of steel.

Basic lead silicochromate, $PbCrO_4.nSiO_2.PbO$
This pigment has been developed comparatively recently and has excellent corrosion-inhibition effects. It is a good deal cheaper than red lead and can be used both in priming and top coatings. The pigment protects underlying steelwork by a combination of alkaline character and oxidising effect. One of the biggest contracts on which this pigment has been used recently is the Verrazano Narrows bridge near New York City.

Zinc and barium–potassium chromates
Zinc chromate is used mainly for priming coats for magnesium/aluminium alloys, because lead compounds attack these metals. One has to be rather careful as to the kind of medium one uses with zinc chromate, as the pigment reacts with many of them. The best to use are phenolic types. Zinc chromate is unsuitable for use with stoving enamels as it breaks down at temperatures above about 170°C. It is also unsuitable for use in acid environments or even in highly polluted atmospheres. Barium/potassium chromate, which is made by heating potassium dichromate, chromic acid and barium carbonate at about 650°C, can be used both for stoving enamels and also in acid atmospheres. The pigment is most suitable for finishes intended for light alloy surfaces.

Zinc polymolybdate, $5ZnO.7MoO_3$
Zinc polymolybdate is used in so-called long-oil alkyd coatings and has been found to have excellent anticorrosion properties. These have been found to be superior to those of red lead. Calcium molybdate is also used for such purposes.

METAL POWDER PIGMENTS

Three metal powders are incorporated into surface coatings: aluminium, zinc and lead.

Aluminium powder

Aluminium powder is manufactured by stamping out tiny leaves of aluminium flakes from thin aluminium sheeting. The leaves are lubricated with stearic acid and suspended in this form in a paint medium. Non-leafing aluminium powder is also used but is less popular. Various flake sizes are made, the smaller sized ones being better from the point of view of surface finish. Aluminium paint intended for surface protection usually contains about 20% by weight of aluminium powder. Aluminium pigment protects against corrosion mainly by virtue of the fact that a finish incorporating aluminium powder is extremely impermeable to moisture and able to exclude ultra-violet light. The leaf-like pigment particles settle in a horizontal position on the outside of the paint film. This protects the paint medium from light, and thus prevents it from being broken up as may happen with inadequately pigmented media.

The aluminium flakes are oxidised normally, and thus do not have any anodic protective effect, except against chloride ions which penetrate the oxide film. Aluminium foil suspended in either silicon or silicon-alkyd paints are used as high-temperature finishes for boilers, exhaust pipes and other units subjected to constant or intermittent heating.

Zinc powder

Zinc powder used for pigment purposes has a particle size of between 3 and 10 μm. When formulated in paints it must be combined with a suspending agent as it is otherwise liable to settle down. Paints that are used for corrosion-resistance purposes contain about 20% by weight of zinc. Other pigments such as iron oxide can also be present. The protective action of zinc powder is due to its anodic electrochemical position against iron and steel. Its effect is very similar to that of a galvanised coating. Zinc powder is used mainly in rubber base, polyurethane, epoxy, silicon and alkyd-based media, and the paints are used to protect large steel structures like ships, bridges, cars and overland pipelines. Zinc powder containing some metallic cadmium is also used for protective coatings.

Lead powder

Finely divided lead is dispersed in a paint medium in conjunction with extenders such as asbestine, barytes and basic lead sulphate. Such a mixture can have a lower density than red lead paint and is found to be extremely effective as a rust inhibitor. The main reason why lead powder is effective as a rust inhibitor is that it acts as a scavenger for oxygen.

8.7 Application of paints

It is absolutely essential that paints are applied in such a way that films are completely continuous and remain so—the existence of even tiny pinholes may cause the paint film to lift, thereby causing rapid corrosion. It is essential to obtain good adhesion of the paint to the underlying surface and to avoid blistering, which may take place when salt ions pass through the film and attack the underlying metal. Gases are then evolved which lift off the film. Polyurethane, vinyl and epoxy–polyamide coatings resist blistering well even after prolonged immersion in salt water. Alkyds, phenolics and epoxy–amine coatings perform badly in this respect.

Paints are normally applied by brushing, rollers, spraying or dipping. In all cases it is essential that the solvent balance and the viscosity are controlled rigidly and accord with the maker's specifications. It is vital that during application of the paint no uncovered patches are left, as these, which usually tend to have a less ready access to oxygen as well, corrode rapidly due to a combination of differential aeration and crevice corrosion. In the case of sprayed finishes, this local corrosion can often be avoided by the use of electrostatic spray methods in which the paint particles are specially attracted to hidden portions. Dipping too has the advantage that hidden parts are covered more thickly by paint than flat surfaces, because of the effect of capillarity which applies with crevices. Paint coatings applied by brush and roller tend to be thicker than those applied by spraying or by dipping, particularly if a thixotropic paint is used.

SURFACE PREPARATION OF METAL OBJECTS PRIOR TO PAINTING

Steel surfaces

Steel surfaces are among the most difficult to prepare properly for the application of paints and other surface finishes. In most cases steel that has been exposed to the atmosphere for some time is covered by coatings of rust, which is inevitably contaminated with soluble salts. The most common of these are ferrous sulphate and ferric chloride. They exert an osmotic effect, which is of a very high order of magnitude. Moisture on the surface of the paint is forced through the paint film and serves to corrode the steel underneath. Steel that has been heat treated is covered with a film of mill scale which also tends to carry soluble iron salts within its pores. It therefore becomes necessary to ensure that every trace of rust is removed from the surface of a piece of steel prior to painting. The answer to the problems involved is, however, not as simple as this, because really adequate surface preparation of steel is very expensive, and often a method which may cost very much more, has only a fractionally beneficial effect on the durability of the paint applied.

For objects small enough to be treated in a factory, the best method would be the removal of scale by means of sulphuric acid containing phosphoric acid as an inhibitor followed by immersion in phosphoric acid, washing and drying. Alternatively, one of the various conversion coatings which acts as a good base for the application of paint can be used.

For structures that have already been erected, such as pylons, bridges and gasometers, the question of surface preparation is more complex.

The most effective, but also the most expensive, method is sand blasting. Flame cleaning is slightly less effective, but only costs about 75% as much as sand blasting. Hand and power wire brushing costs between 10 and 20% as much as sand blasting, but is far less effective.

When repainting structures it has been found to be most economical to start repainting before undue damage of the priming coat has taken place, since the removal of rust once this has set in, is a very expensive matter.

Aluminium, magnesium and other light alloys
When applying paints to aluminium surfaces it is best to produce an anodised surface, as paint adheres very much better to it than to untreated metal. When painting aluminium structures that are likely to be exposed to external conditions, a priming coat containing zinc chromate is found to be of value.

Copper, brass and zinc
Copper and brass are commonly treated with a 20% sodium dichromate solution containing some free sulphuric acid. This produces a conversion coating on their surfaces which has much improved adhesion to stoving enamels and other surface coatings. Zinc, whether in the form of the metal itself or galvanised coatings, is treated with phosphoric acid or chromic acid to produce a suitable conversion coating with the necessary adhesion to paint. It has even been found useful to wait some six months before painting galvanised steel structures, as paint adheres better to somewhat weathered zinc surfaces than to fresh and shiny ones.

8.8 Hot organic coatings

Surface coatings which are applied hot include the following:

1 Asphalts and related materials.
2 Synthetic hydrocarbon resins.
3 Cellulose derivatives.
4 Waxes of all types.

ASPHALTS AND RELATED MATERIALS

Natural asphalt, which is found in the USA, Trinidad and Venezuela, is a black material with a density between 0·9 and 1·0 kg/dm^3 and a melting-point of between 76 and 86°C. It contains a mixture of wax esters, wax acids and resins, and has an average molecular weight of about 1,500.

Coal tar pitches have a higher density, which varies between 1·4 and 1·5 kg/dm^3. The softening point of coal tar pitches is of the same order as that of natural asphalts.

Both are widely used for the coating of submerged pipelines, for the production of roofing felt, and surface coating of concrete and other structures permanently exposed to water. Asphalts have excellent resistance to most non-oxidising acids but are attacked almost immediately by any non-polar solvent. Their resistance to alkalis is good and to inorganic salts excellent.

SYNTHETIC HYDROCARBON RESINS

The main resins used for hot surface coatings are:

1 Resins are made from the coumarone–indene fraction which occurs in the light oil of coal tar distillates. These are used for chipboard coatings, adhesives for bonding aluminium sheeting to wood and similar purposes.
2 Terpene resins which have a molecular weights between 650 and 1,600.
3 Polyisobutylene and cyclised rubbers, which are employed as hot coatings for materials of a pliable nature.
4 Polyethylene is applied hot and also has the property of good flexibility. The grades used for this purpose have molecular weights which vary from 1,000 to 9,000.
5 Polystyrene is blended with various waxes to produce hot applied coatings with high gloss, good elongation and excellent adhesion.
6 Resins incorporating chlorinated hydrocarbons, such as chlorinated biphenyl, terphenyl and napthalene, are blended with waxes to give flame-resistant coatings. The material has the trade name 'Hypalon' and is cured after application to surfaces.

CELLULOSE DERIVATIVES

The main cellulose derivatives used for hot coatings are cellulose acetate–butyrate resins plasticised with dibutyl and dioctyl sebacate. These materials have melting-points of between 120 and 140°C depending on the nature and quantity of plasticiser added. The material is widely used for hot dipping purposes, mainly for the surface protection of smaller parts.

WAXES OF ALL TYPES

Waxes can be subdivided into vegetable, insect and mineral types. The composition of waxes varies very widely. Table 8.1 summarises the main properties of commercial waxes.

TABLE 8.1

Wax	Melting-point, °C	Density, kg/dm³
Mineral wax, $C_{42}H_{20}$	37	0·777
Mineral wax, $C_{54}H_{26}$	56	0·778
Mineral wax, $C_{62}H_{30}$	66	0·780
Chlorinated paraffin wax, 70% Cl	86	1·64
Shellac wax	82	0·93
Bee's wax	50	0·97
Spermaceti wax	45	0·96
Carnaba wax	84	0·999
Candelilla wax	67–79	0·947
Sugar cane wax	80	0·983

Waxes are used both on their own and together with various resins as hot coatings to protect metals.

Literature sources and suggested further reading

1 FISCHER, W. VON, *Paint and varnish technology*, Hafner, New York (1964)
2 CHATFIELD, H. W., *The science of surface coatings*, Benn, London (1962)
3 *Paint technology manuals*, 6 volumes, Chapman and Hall, London (1960–1966)
4 TATTON, W. H., and ANDREWS, E. W., *Industrial paint application*, Newnes, London (1964)
5 BAKHALOV, G. T., and TURKOVSKAYA, A. V., *Corrosion and protection of metals*, Pergamon Press, Oxford (1965)
6 BURNS, R. M., and BRADLEY, W. W., *Protective coatings for metals*, Reinhold, New York (1967)
7 FONTANA, M. G., and GREENE, N. D., *Corrosion engineering*, McGraw-Hill, New York (1967)
8 CHANDLER, K. A., 'Paint failures', *Building maintenance* (July 1969)
9 *Electrophoretic paint deposition*, R. H. Chandler, London (1970)
10 *Proceedings of the Third International Congress on Metallic Corrosion*, Sweys and Zeitlingen, Amsterdam (1970)
11 GREENWOOD, J. D., *Heavy chemical and electro-deposition*, Robert Draper, Teddington (1970)
12 SPELLER, F. N., *Corrosion, causes and prevention*, McGraw-Hill, New York (1951)

Chapter 9   Miscellaneous Corrosion-resistant Coatings

9.1 Chemical conversion coatings

Chemical conversion coatings on the surfaces of metals are produced by treating the metal chemically in such a way that an appropriate compound is formed. In general, such coatings when applied to iron or steel are not good enough to protect on their own. However, they form an extremely good basis for the application of paint films or plastic coatings.

PHOSPHATE COATINGS

Phosphate coatings are normally applied by immersing the steel object in a bath containing equal weights of zinc phosphate $Zn_3(PO_4)_2$ and phosphoric acid H_3PO_4. Under such conditions a film of monohydrogen phosphate is formed on the object as follows:

$$Fe + H_3PO_4 \rightarrow FeHPO_4 + H_2$$

Zinc phosphate is also precipitated on the surface of the iron by somewhat complicated secondary reactions. The reaction is usually speeded up by adding oxidising agents such as nitrates, and the quality of the phosphate coating is improved by the presence of small quantities of manganese.

Phosphate coatings are modified for specific purposes. To produce a black surface one adds either zinc sulphide or calcium nitrate to the solution. A cheaper phosphate coating, which is used when paint is applied immediately afterwards, is obtained by the use of a mixture of sodium phosphate, phosphoric acid and detergent.

Phosphate coatings on their own have considerable resistance to corrosion, the rate of attack usually being about 10% of that for untreated surfaces under the same conditions. The weight of phosphate applied in standard treatment baths is usually rather low. On average, the weight of phosphate per square metre amounts to between 1 and 2 g. The thicker the coatings the greater their protective nature but such coatings are seldom used on their own, because they can readily be punctured and then accelerate corrosion of the steel surface. Their porous nature makes them extremely valuable as a base surface for the application of all kinds of paints and other organic coatings. It has been found that the rate of corrosion is very much lower when paints are applied on phosphate coatings than when they are applied to surfaces which have been sandblasted or merely degreased.

Phosphate coatings tend to absorb oil and wax and therefore many objects, which cannot readily be painted, are surface treated by phosphating, followed by chromating and then oiling. This method or rust prevention is often used with rifles, machine guns and other weapons of war.

CHROMATE COATINGS

Chromate treatment is commonly used on top of normal phosphated iron and steel objects in order to increase the strength of the phosphate coatings, and to improve their adhesion to paint. Chromating is, however, much more frequently used for the the protection of zinc and cadmium surfaces. Zinc objects, and particularly steel parts that have been galvanised, are immersed in a solution of sodium dichromate, $Na_2Cr_2O_7$, which has been acidified somewhat with sulphuric acid. This forms a film of $Cr_2O_3.nH_2O$ which is only about 0·5 μm thick, on the surface of the zinc. The chromate films can have a variety of different colours, these shades being imparted by the presence in the bath of small quantities of such substances as selenium oxide. Chromate treatment of galvanised sheeting has the effect of very much delaying the first appearance of rust spots on surfaces. Without chromate treatment some galvanised sheeting showed traces of rust after one single day's immersion in stagnant water. Chromate treatment delayed this to 38 days. Similarly the rate of corrosion of chromate-treated galvanised steel sheeting was less than 10% of that of similar untreated galvanised sheeting.

Chromating is also widely used for the protection of aluminium, particularly for aircraft components, and for aluminium objects which are to be coated with paint afterwards. The so-called 'Hinac' process utilises an aqueous solution of a mixture of sodium chromate and a reducing agent such as tannic acid. After application the object is heated to 200°C for some hours. This produces an extremely tough film of $Al_2O_3.Cr_2O_3.nH_2O$ on the surface of the aluminium, which is strong and resilient, and serves as an excellent basis for the application of paints. Coatings applied by this method vary in thickness between 0·01 and 40 μm.

The 'Hinac' process can also be used for conversion coatings on a wide variety of different metals, including magnesium, steel, copper and zinc.

OXIDE COATINGS

Coloured finishes are applied on surfaces of iron and steel by oxidation using various immersion and heating techniques. A strongly alkaline solution containing a mixture of sodium nitrite, sodium nitrate, sodium chlorate and free sodium hydroxide is used for the production of a black oxide film on the surface of steel objects. The bath also contains small quantities of the sodium salts of tannic and tartaric acids, and is normally operated at about 300°C. Immersion usually takes about 10 min. At the end of that time the object is coated with a film of Fe_3O_4, between 0·8 and 1·8 μm thick. Such a coating is abrasion-resistant and can be used with stressed objects, since any possibility of hydrogen embrittlement is absent.

If steel is immersed in a bath containing a mixture of molten sodium and potassium nitrate, or heated in a furnace to about 400°C, it is covered on its surface by a blue oxide film.

Stainless steel is given a black oxide film surface by immersion in molten sodium dichromate at 400°C.

Oxide films of various colours can also be obtained by direct heating. A deep slate-grey colour is produced by heating steel objects to red heat in an oven, followed by exposure to superheated steam. Thick protective oxide coatings are also obtained by heating steel articles in air to about 500°C, exposing them first to steam and finally dipping them into well-oxidised linseed oil.

Although oxide coatings on their own have a slightly better corrosion resistance than phosphates, they are usually protected by oiling, waxing or painting. Like phosphate coatings, oxides offer an excellent base for such media.

OXALATE COATINGS

A recently developed method for producing integral conversion coatings on steel, alloy steel and even stainless steel, is the use of solutions of oxalic acid, together with traces of sodium chloride and sodium thiocyanate. The oxalate coatings produced are used particularly to protect steel from attack during mechanical working processes.

ANODISING

Anodising is the formation of a relatively thick and coherent oxide film on the surface of a metal by making the metal the anode of an electrolytic cell and passing a current.

The most common electrolytes used are chromic acid, sulphuric acid, oxalic acid, phosphoric acid, and mixtures of these. Anodic coatings formed in baths of these materials tend to be porous but constitute excellent surfaces for the application of paint.

A very hard and completely non-porous film is produced by the use of a mixture of boric acid and sodium borate. The thickness of film that can be applied is directly proportional to the voltage used. In general one can reckon upon a thickness of 1·5 nm for each volt applied. This means that the deposition of a coating, 1 μm thick needs about 660 V.

A number of trade names are concerned with anodised coatings: 'Eloxal' baths contain oxalic acid, while 'Aluminite' baths contain sulphuric acid. Sealing of anodised films increases their strength and corrosion resistance. To seal anodised coatings, they are afterwards subjected to heating in either hot water or steam, which modifies the crystalline structure of the aluminium oxide on the surface of the object. Immersion in potassium dichromate serves the same purpose. Often anodised coatings are coloured, the dye usually being applied before the sealing process. The dyeing materials are usually organic in nature, but are often inorganic hydroxides, which are co-precipitated during the anodising process. Copper, nickel, cobalt and chromium hydroxides are frequently employed for this purpose.

Bright anodising is a process that combines polishing of aluminium metal, followed by the protection of this shiny surface using anodic deposition of a minutely thin film of aluminium oxide. Bright anodising is best used with quality alloys which contain some magnesium since it is necessary that not even traces of iron, silicon or manganese be present in the alloy. The metal is exposed to a mixture of phosphoric and nitric acids, containing some acetic acid as well. Careful adjustment of the e.m.f. produces 'electropolishing', i.e. the removal of a very thin film of metal from its surface together with all surface impurities. The aluminium object which now has a very shiny surface, produced by the electropolishing process, is then washed and transferred immediately to a bath of sulphuric acid where it once more forms the anode. A thin film of aluminium oxide is formed on the surface, which in no way detracts from the shiny appearance of the metal. The reflectivity of anodised shiny aluminium is maintained even on prolonged exposure to polluted atmospheres. It has been found that after one year's exposure to a heavily polluted atmosphere the total reflectivity of such surfaces is still about 0·83. A polished, but unanodised sample of aluminium would have a reflectivity of only about 0·72 after the same period and conditions of exposure.

Anodic coatings which are produced with the help of a chromic acid bath tend to be the thickest, being up to 12 μm thick, and thus stand up best to corrosion. Coatings which are applied using the sulphuric acid treatment tend to be thinner, namely only 6 μm, but are tougher, particularly after the hot water, steam or dichromate sealing treatment. Anodising with the same electrolyte solutions as are used for aluminium, also gives protection to the following metals: beryllium, copper, magnesium, nobium, tantalum, titanium, zinc and zirconium.

9.2 Temporary protection for metal surfaces

Metal objects can be protected temporarily against corrosion by two methods:

1 By the use of a temporary plastic or rubber film, which can be readily removed when required.
2 By painting the surfaces with oil, grease or other materials capable of removal.

STRIPPABLE FILMS

The best known of these methods is the so called 'cocooning', which is widely applied for the protection of aircraft engines and similar equipment during transit. The equipment is first of all taped, so that the plastic material is kept away from vital components. Then, using a spray gun, solutions of vinyl resins or cellulose acetobutyrate are sprayed in such a way that a thin web is formed around the equipment. Further coatings are then sprayed until the cocoon is totally impervious to water vapour and acid gases. A hole is cut into the cocoon and hot air is pumped through, to remove any traces of moisture present. Silica gel in trays is also inserted, to remove traces of water vapour which may accidentally penetrate the cocoon protection, and the slit through which the trays were inserted is carefully sealed.

Strippable surface coatings of plastic are often used to protect metal objects during storage and transportation. Vinyl coatings between 0·03 and 0·1 mm thick are applied by means of a spray, and similar coatings of ethyl cellulose are applied by dipping. A method which is becoming increasingly important is that of coating steel objects with a vinyl gel, which is cured at 230°C for 90 sec using infra-red radiation. This forms a thick film which contains non-interconnecting pores. Rubber latex is also often employed as a strippable film, but it is necessary to use it in conjunction with a sodium benzoate or sodium nitrite inhibitor, otherwise the sulphur and chlorine contained in the latex causes attack on iron or steel surfaces.

GREASING OR WAXING OF SURFACES

Oils and greases used for rust protection normally need a corrosion inhibitor in their formulations and in addition it is necessary that they themselves are entirely non-corrosive, i.e. that they do not contain appreciable acidity. Mineral oils are not too good, because they have no active groups which attach themselves to the metal they cover. It is necessary to add activated groups in the form of sodium salts of sulphonic organic acids, or calcium salts of long-chain aromatic compounds such as calcium stearate. Lead stearates, oleates, etc., are also widely used and have the additional advantage of some corrosion protection.

Oils and greases are usually applied in the form of a solution in either coal tar naphthe or petroleum distillate (white spirit). Steel sheeting, covered with lanolin by brushing, spraying or dipping at a rate of 2 g/m^2, is protected against corrosion in a normal indoor atmosphere for several years.

Bituminous rust preventatives are also often used, again dissolved in solvents. This technique shows a slight advantage over the lanolin treatment but costs between 3 and 4 times as much to carry out.

Wax films are applied by dipping the object into a bath containing molten wax and withdrawing. Alternatively the wax can be brushed or even sprayed on as a solution in a solvent. Wax films are much thicker than either lanolin or bituminous films and, in consequence, they give very much better protection. Wax films completely protect steel objects for three years indoors and up to two years outside. Wax coatings are the only satisfactory method for this preservation of such highly finished parts as razor blades, ball bearings and inlet and exhaust valves for motor cars. Coatings applied are usually about 50 μm thick.

Preservative oils are generally applied in much thinner layers than greases. In consequence these coatings, which seldom exceed 6 μm in thickness, are only commonly used to protect steel parts in between manufacturing operations, or finished parts which have been treated by a chemical

N

conversion coating. The oils used are specially formulated to adhere to metal surfaces even if they are damp, and also include rust-inhibiting chemicals.

9.3 Plastics and rubber coatings

Paint films (see Chapter 8) are generally less than 0·2 mm thick. The plastic coatings described here are much thicker and in consequence the protection given to the underlying metal is much better. Thick plastic coatings, or mastics as they are often called, are used when steel objects are in permanent contact with corrosive fluids. The most common use for rubber mastics is in the undersealing of motor vehicles.

Plastics used for thick coatings are epoxy resins, fluorinated hydrocarbons, polyesters, polyethylene, phenolics, silicones and vinyl compounds. These are applied by a variety of different methods:

1 Thermoplastic resins such as polyethylene or vinyl materials are prepared in the molten state and objects are dipped into the tank containing the melt. The thickness of film applied is controlled by the viscosity of the melt. Centrifuging is sometimes used.

2 True two-pot resins such as epoxies are applied by mixing the resin with its catalyst and applying by brush or trowel. Often sheets of plastic are glued on, using catalytically setting cements of this type for the purpose.

3 Thermosetting resins such as phenolics are dissolved in a hydrocarbon solvent and sprayed on to the surface of the object. After evaporation of the solvent, the plastic is cured by infra-red heating.

Really thick plastic linings—needed for plant used in the chemical industry—are often made by laying several coatings of plastic on top of each other. Such coatings may be up to 10 mm thick.

PVC COATINGS

PVC or polyvinyl chloride is one of the most widely used plastics in the field of thick coatings. PVC is rather hard when used in the unplasticised state, but then shows excellent resistance to many chemicals. Although the addition of plasticisers reduces this resistance somewhat, adhesion to base metals is very much improved by the addition of materials such as tricresyl phosphate and linseed oil. PVC coatings are generally applied by dipping objects into a plastisol mix of PVC, followed by curing at about 180°C. These coats are normally between 1 and 1·5 mm thick. Materials that are becoming of increasing importance in the building and construction industries are steel and aluminium sheeting, which are bonded to so-called duplex vinyl sheeting. The vinyl coating material is made in the form of two layers, the top layer which faces outside being formulated from non-plasticised vinyl which has excellent resistance to chemicals, while the bottom layer, which is bonded to the steel or aluminium, is heavily plasticised. This sheet is then glued or cured against the surface of the metal base. In the case of aluminium, it is necessary to anodise the surface of the metal to enable the vinyl sheet to stick. The coatings of vinyl are very thick and extremely abrasion resistant. They adhere firmly even if the base metal is severely deformed by pressing, etc.

POLYETHYLENE-BASED COATINGS

Polyethylene is a plastic material that resists most acids and bases, including even nitric and hydrofluoric acids. It is basically a paraffin hydrocarbon with an enormous chain length (between 50,000 and 150,000 carbon atoms). The longer chain lengths of polyethylene have a higher density, which means that they have rather better chemical resistance and also better temperature stability. Low-density polyethylene is easier to apply to metals. Metal surfaces are usually primed and the plastic is attached using dusting, hot dipping, or alternatively, dissolving the polyethylene in solvents such as trichlorethylene and spraying on. In general, adhesion to hot surfaces is a good deal better than to cold surfaces. Polyethylene films used in industry vary between 0·3 and 3 mm in thickness.

Polyethylene linings are widely used for industrial vessels such as tanks and pipelines and also for such purposes as the protection of the internal surfaces of tin cans containing acidic liquids such as fruit juices.

Closely related to polyethylene is Hypalon, which is chlorosulphonated polyethylene. This material has excellent resistance to many chemicals which attack normal polyethylene, e.g. oils, ozone and strongly oxidising salts.

PTFE (polytetrafluoroethylene) is a plastic material that resists all common chemicals with the exception of molten sodium. In addition, it can be heated to 250°C without any adverse effect. Application is by the use of dispersions which are sprayed or brushed on, followed by heating. PTFE coatings are commonly about 50 μm thick.

RUBBER COATINGS

Most forms of rubber bond readily to metals with considerable tenacity. Such bonds are due to direct chemical forces. When vulcanised rubber is bonded to copper, for example, a reaction takes place between the sulphur of the rubber and the copper producing coordinated S–Cu bonding (see Fig. 9.1). When rubber is to be bonded to metals other than copper, it is usually necessary to apply a layer of ebonite to the surface of the metal by means of a roller, or in the form of a solution. Rubber is attached to the ebonite by cross-linking. The strength of such metal–rubber bonds is unaffected by low temperatures, but is weakened when temperatures are increased. Chlorinated rubber, produced by the reaction of hydrogen chloride with rubber dissolved in benzene, also bonds readily to many metals. The chlorine content of this form of rubber may be as high as 68%. It is used frequently as an intermediate layer to permit neoprene and nitrile rubbers to be attached to metals. Chlorinated rubber cannot be used directly with natural or styrene–butadiene rubbers, because of incompatibility. Interlayers of neoprene are required in such cases.

Rubber sheeting is often cemented directly to metal surfaces. Rubber has excellent chemical resistance to many reagents. It withstands any concentration of hydrochloric, hydrofluoric and most organic acids, any concentration of caustic alkalis, acetone or alcohol and most non-oxidising inorganic salts. Rubber also resists sulphuric acid up to 75% in strength. In general the maximum temperature recommended is 65°C, although in contact with oxidising agents the maximum temperature to which rubber films may be subjected is a good deal lower than this.

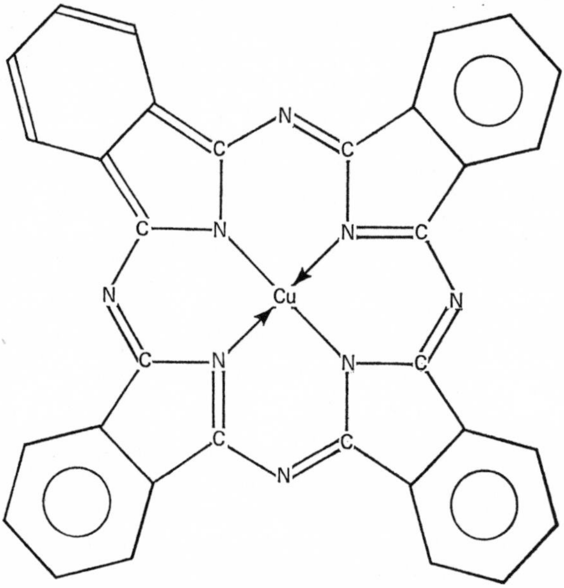

Copper phthalocyanine

Fig. 9.1. Structure of copper phthalocyanine

9.4 Bitumen (asphalt)

Bitumen, or asphalt as it is called in the United States, is the residue obtained from the distillation of crude oil. In Great Britain the term asphalt is usually reserved for mixtures of bitumen with inert mineral matter. Some bitumen is also found naturally either alone or mixed with mineral matter. The material has a variable hardness depending upon the quantity of light fractions removed during the distillation process. Its softening point is between 25 and 170°C, depending on grade.

Bitumen can be oxidised by bubbling air through the residue during manufacture. This produces a product that has some rubbery properties and which does not soften as readily on heating or become as brittle on cooling as ordinary grades of bitumen. Bitumen is used as an anticorrosive coating and is usually applied by dissolving the molten substance in a solvent in a closed vessel. These coatings are then applied by brush and spray. Two coats will protect steel structures against corrosion in an industrial atmosphere for about one year. Thicker coatings of bitumen (in excess of 0·5 mm) are usually applied by dipping or trowelling of the molten bitumen itself. Bitumen has excellent chemical resistance to acids and bases, as well as to inorganic salts. The resistance of bitumen to organic solvents is virtually nil. To increase the strength of bituminous finishes they are often coordinated with resins such as phenolics, alkyds, urea formaldehyde or polystyrene. Such finishes are used for treating the external surfaces of pipelines, employed to carry water, gas or oil.

9.5 Ceramic coatings

These are inorganic coatings using mainly a silica base, which are attached chemically to the metal surface. The advantages of ceramic coatings are that they are generally able to withstand temperatures up to 500°C, are resistant to abrasion and attractive in appearance. Their disadvantages are their liability to crack under conditions of mechanical and thermal shock and the cost of applying them.

PORCELAIN ENAMELS (VITREOUS ENAMELS)

The constituents of these enamels somewhat resemble those used for the manufacture of borosilicate glasses. The formulations include mixtures of B_2O_3, SiO_2, Al_2O_3, ZrO_2, Na_2CO_3 and PbO. The exact composition used varies widely. If the steel is to be used in acidic atmospheres or in contact with acids, the percentage of SiO_2 is increased, and porcelain enamels for use in conjunction with basic materials include a higher proportion of Al_2O_3 and ZrO_2.

The glass constituents are melted to form a clear mix. After having been cooled, the glass is ground up to form a material which is called 'frit'. This is mixed with a number of other materials such as CaF_2, $NaNO_3$ and $CaCO_3$.

Vitreous enamelling is carried out in two ways:

1 The wet process.
2 The dry process.

In the *wet process* the powdered frit is suspended in water together with clay, and the mix which contains about 15% solids, is sprayed on to the base metal to be treated. The object is then heated gently to dry the coating, followed by firing to 850°C for a period of about 5 min, or less if the objects are made from light-gauge steel sheeting. The process is repeated several times until the required thickness of ceramic coating has been achieved. The nature of the frit mix for the different coats varies. The undercoating contains quantities of Co_2O_3 and Ni_2O_3, as it has been found that this produces good adherence to the steel surface. The top coating contains various metallic oxides whose function is to provide the necessary colour to the enamel finish.

The following oxides are used:

Opaque white	TiO_2
Red	CdO, Fe_2O_3 or SeO
Blue	Co_2O_3
Green	Cr_2O_3
Yellow	Sb_2O_3

Apart from spraying, it is possible to use dipping and even flow coating for the application of the suspended frit. The wet process of vitreous enamelling is widely used for the production of road signs and enamelled commercial shields.

In the *dry process*, the object to be treated is heated to the fusion temperature of the frit, and the powdered frit is dusted over the surface. The temperature to which the base metal is heated depends on the chemical nature of the frit employed. For iron and steel, firing temperatures of around 700°C are used. Special low-melting-point vitreous enamels have been developed for coating aluminium and aluminium alloy castings. These operate with a firing temperature as low as 500°C, which is well below the melting-point of pure aluminium (660°C). The object is then returned to the furnace where the heating temperature is carefully controlled. The process is repeated until the ceramic coating has the required thickness. Dry application of the procelain coating produces very firm adhesion of the coating to the base metal. For this reason this technique is better than the wet method as regards corrosion resistance. It is widely used for the coating of cast iron objects, for which a frit rich in PbO is used to lower the firing temperature.

The thickness of vitreous enamel applied per coating stage is the same whether dry or wet coating is employed and usually amounts to between 0·08 and 0·12 mm. It is possible to produce thicknesses as thin as 0·03 mm and as thick as 0·05 mm using vitreous enamelling techniques. Thinner films are more flexible and are less liable to be damaged by mechanical or thermal shock. Thick films on the other hand have a longer life when exposed to corrosive media and do not suffer from porosity.

Vitreous enamel can be used with any base metal, provided that the melting-point is high enough not to be affected by the required firing treatment. This means that the method is unsuitable for low-melting-point eutectics of magnesium and aluminium, but can be used both for the pure metals and their high-melting-point eutectics. Steels which are to be vitreous enamelled should have either a very low carbon content or, alternatively, should contain up to 0·3% titanium as a scavenging agent. Free carbon on the surface tends to oxidise during the firing process, causing bubbles and other unevenesses to appear in the vitreous enamel film. Highly refined iron with a carbon content of below 0·06% is very suitable for vitreous enamelling with a high-melting-point frit. Grey cast iron, which contains a high percentage of carbon, can only be vitreous enamelled satisfactorily with a low-melting-point frit. The reason is again the liability of carbon particles on the surface of the metal oxidising to CO and CO_2 gases if a high firing temperature is used. Iron and steel surfaces for treatment by vitreous enamelling are degreased, pickled and sand-blasted. Aluminium and magnesium surfaces are also surface oxidised by anodic treatment in a chromate bath in order to produce better adhesion of the enamel to the metals.

CHEMICAL AND OTHER RESISTANCES OF PORCELAIN ENAMEL

Porcelain enamel on the surface of metals is virtually a glass coating with basically the same resistance to chemical agents as glass. This means that vitreous enamels have excellent corrosion resistance to most acids and salts, but are attacked by hydrofluoric acid and alkalis. Resistance of porcelain enamels to normal exposure of even highly polluted atmospheres is virtually infinite, provided the coating remains intact and free from cracks. The resistance of ceramic coatings to mechanical and thermal shock is markedly dependent upon the thickness of the coating. The best coatings from this point of view are very thin ones, applied in one single operation. Some of these modern 'one-coat' enamels are very elastic and even permit the drilling and cutting of the base metal, without cracking. Similarly, coatings below 0·05 thick can generally be quenched suddenly from 200°C without cracking. Ceramic coatings should be used at a minimum of 200°C below their firing temperatures.

HIGH-TEMPERATURE CERAMIC COATINGS

These are coatings for metals and are able to withstand working temperatures up to 1,700°C.

Normal steels, when protected by ceramic coatings, can be used up to a temperature of around 1,000°C; even aluminium can be used up to 900°C if suitably protected by a ceramic coating. High-temperature ceramic coatings are always applied by means of a wet technique, and are fired to thicknesses of around 40 μm. The composition of the frit used varies extensively. A typical formulation, given as percentages by weight is the following:

Silica	16
Barium carbonate	23
Boric acid	5
Calcium carbonate	3
Beryllium oxide	1
Chromic oxide	17
Zinc oxide	2
Clay (suspending agent)	3
Water	30
	100

A mix rather similar to this but without the boric acid is used for coating stainless steel and Inconel used in nuclear reactors. These can be used for temperatures up to 1,100°C.

Really high temperatures can be withstood by molybdenum, niobium, tantalum and tungsten when they are coated with special ceramic materials. Molybdenum is coated using mostly pure SiO_2 in the coating slurry. When fired at a high temperature this combines with molybdenum base to form $MoSi_2$ on the surface. It is possible to expose molybdenum which has been treated in this way to temperatures of up to 1,650°C without the base metal being attacked. For niobium also, coating is carried out by means of silicon, which reacts with the metal to form extremely stable niobium silicide. This withstands temperatures up to 1,650°C and is used in aircraft jet engines.

Apart from silica, a number of other oxides are used for high temperature purposes. Alumina (Al_2O_3) is widely used because it has good resistance to abrasion and thermal shock. Zirconia (ZrO_2) is used where it is necessary for the ceramic to have low thermal conductivity. Beryllia (BeO) is used where high electrical resistance is required. In addition, a number of other oxides are used either alone or combined with the ones given above, where special requirements exist. Up to now use has been made of the following: Cr_2O_3, HfO_2, MgO, ThO_2, TiO_2 and Y_2O_3.

Literature sources and suggested further reading

1 *Proceedings of the Third International Congress on Metallic Corrosion*, Sweys and Zeitlingen, Amsterdam (1970)
2 GREENWOOD, J. D., *Heavy chemical and electro-deposition*, Robert Draper, Teddington (1970)
3 BONNER, P. E., and WATKINS, K. O., *The priming of sprayed aluminium and zinc coatings on steel*, BISRA, London (1969)
4 BAKHALOV, G. T., and TURKOVSKAYA, A. V., *Corrosion and protection of metals*, Pergamon Press, Oxford (1965)
5 BURNS, R.M., and BRADLEY, W. W., *Protective coatings for metals*, Reinhold, New York (1967)
6 *Proceedings of the First International Congress on Metallic Corrosion*, Butterworths, London (1962)
7 FONTANA, M. G., and GREENE, N. D., *Corrosion engineering*, McGraw-Hill, New York (1967)
8 NYLEN, P., and SUNDERLAND, E., *Modern surface coatings*, Interscience, New York (1965)
9 PAYNE, H. F., *Organic coating technology*, 2 volumes, Wiley, New York (1954)

Chapter 10   **Corrosion of Boiler Plant**

When studying the corrosion phenomena that occur in a boiler installation, it is necessary to distinguish between two entirely separate aspects. These are the corrosion processes that take place on the 'water' side of the boiler, i.e. inside the boiler tubes, the superheater, economiser and condensing equipment, and the corrosion on the outside of these surfaces, i.e. on the 'fire' side of the equipment.

Figure 10.1 shows a diagrammatic section of a modern tube-type boiler installation and the relative positions of such features as the economiser, superheater, waste heat boiler, etc. High-pressure boilers which drive turbines operate on a recirculatory system, and only make use of so-called 'make-up' quantities of water, which are very carefully treated by means

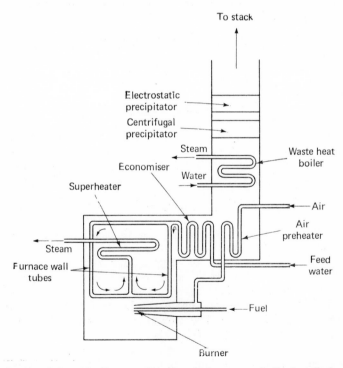

Fig. 10.1. Schematic diagram of modern high-pressure boiler installation

of modern ion-exchange methods. Low-pressure boilers, particularly those that provide either hot water or process steam, use less well treated water. Water in such a case is either totally untreated or is softened by the use of chemicals such as sodium aluminate. Magnetic water treatment is also becoming more popular since this flocculates the deposited material and avoids scale deposition.

10.1 Corrosion on the water side of a boiler

Water used for steam raising often contains dissolved solid and gaseous impurities. These can cause scaling and corrosion (which are often inter-related) in the boiler plant. In modern high-pressure boilers the solid content can be lowered to negligible proportions by the use of water-treatment methods such as mixed-bed ion exchange. Water purities as high as 99·999997% have been quoted.

ACTION OF WATER ON MILD STEEL INSIDE BOILERS

At temperatures above 250°C water attacks steel to form a surface layer of magnetite Fe_3O_4. The reactions are electrochemical and take place as follows: at the anode

$$3Fe - 8e \rightarrow Fe^{2+} + 2Fe^{3+}$$

at the cathode

$$8H^+ + 8e \rightarrow 4H_2$$

The total reaction is

$$3Fe + 4H_2O \rightarrow Fe_3O_4 + 4H_2$$

The exposed iron is the anode, while the surface underneath the magnetite film acts as the cathode. All this produces a coherent film on all internal surfaces of steam boiling equipment. In the presence of dissolved oxygen iron is oxidised to porous Fe_2O_3 instead of non-porous Fe_3O_4:

$$4Fe + 2nH_2O + 3O_2 \rightarrow 2Fe_2O_3 . nH_2O$$

The areas shielded from oxygen supply, i.e. the crevices, are preferentially attacked, causing bad pitting.

Under normal circumstances the magnetite film on a new boiler after conditioning is about 0·025 mm thick, but may reach a thickness of 0·25 mm after ten years' use at 100 bar pressure. The coherent and protective magnetite film is weakened by oxygen in the boiler feed water causing the corrosion of underlying steel. The growth of the protective magnetite film inside a boiler is accelerated by using alkaline boiler water. Excessive local concentration of sodium hydroxide causes the formation of soluble iron ferroates, which weaken the film. Slight acidity retards the formation of the protective magnetite film and excessive acidity removes it. The magnetite film can also be destroyed by either mechanical or thermal shock.

THE EFFECT OF THE pH ON THE BOILER WATER

Acidic boiler water tends to attack the magnetite layer, and the metal surfaces underneath this layer, particularly in places where a crack or joint induces crevice corrosion. The minimum amount of corrosion from this cause is found when the boiler water has a pH as high as 11–12, but this is normally unsatisfactory, because under such conditions the water would be so alkaline that there would be a danger of caustic cracking. Above pH 12, the magnetite becomes very thick and tends to crumble away. In practice, the boiler water is usually maintained at a pH between 8·5 and 9, depending upon the pressure at which the installation is operated.

THE EFFECT OF DISSOLVED OXYGEN

Quite minute quantities of dissolved oxygen cause corrosion in boilers, particularly if these boilers are operated at high temperatures and pressures. As can be seen from the Table 10.1, the higher the operating pressure, the lower the maximum amount of dissolved oxygen permitted. The reason why even very tiny quantities of oxygen in boiler water are so very damaging is that they permit the cathodic part of the corrosion reaction to proceed.

TABLE 10.1   Recommended feed water conditions for boilers operating under varying pressures

Boiler pressure, bar	pH	Dissolved O_2, ppm	Carbonate hardness, ppm	Chlorides, ppm	Total dissolved solids, ppm
15	8·5	0·05	2	3	8
30	8·7	0·03	1	1	3
50	8·8	0·01	Nil	Nil	0·2
100	9	0·007	Nil	Nil	0·05
180	9	0·005	Nil	Nil	0·02

Corrosion processes, like all other chemical reactions are speeded up enormously by elevated temperatures and rates inside boilers would be very fast if the kind of conditions existed inside, as exist under normal external conditions, where ample oxygen is present to allow the cathodic reaction to occur.

Corrosion inside a boiler can only be restricted by keeping the oxygen concentration very low; this stops the cathodic reaction from taking place. Corrosion by dissolved oxygen in boiler feed water proceeds ten times as fast as it would with equal quantities of dissolved carbon dioxide.

OXYGEN REMOVAL

Oxygen is removed from boiler feed water in four distinct ways:

1 Mechanical deaeration methods.
2 Use of hydrazine.
3 Use of sodium sulphite.
4 Ion-exchange techniques.

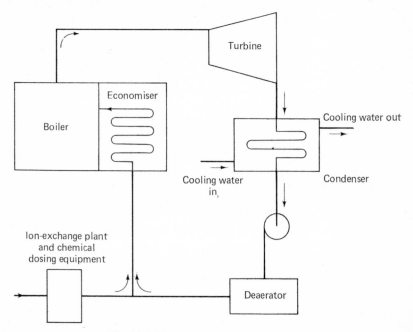

Fig. 10.2. The supply of boiler feedwater

Mechanical deaeration methods

Distillation—Normal distillation of water permits the oxygen concentration to be reduced to 0·1 ppm, but considerable improvements can be made by the use of more modern vapour purification plant. With this plant, oxygen concentrations down to 0·01 ppm can generally be achieved.

Steam scrubbing—There are numerous methods on the market of removing oxygen and other dissolved gases from steam by various 'scrubbing' techniques. Some plants are capable of getting the oxygen content down to below 0·007 ppm, which is the figure usually aimed at in British power-station practice. Where plants are less efficient than that, final reduction of oxygen concentration is carried out by the use of hydrazine.

Desorption—In this process water is mixed with nitrogen that is completely oxygen-free in a jet injector at a pressure exceeding about 3 bar. This technique permits the oxygen concentration to be reduced to about 0·04 ppm, the final reduction to virtually zero being made with hydrazine.

Flash-type deaeration—Pressurised water is heated to a temperature above its boiling point. It is then sprayed into an evacuated chamber (see Fig. 10.3) where it boils, so that most of the incondensible gases can be pumped off. If the feedwater comes from the turbine exhaust, it is possible to reduce the oxygen content to below about 0·07 ppm. Feed water deaerators can reduce the oxygen content to below 0·007 ppm, thereby making it unnecessary to add chemical agents for boilers operating below 100 bar.

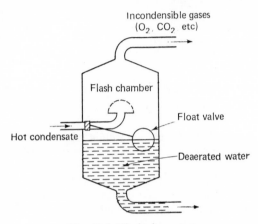

Fig. 10.3. Flash deaerator

The use of hydrazine

The oxygen content of boiler feed water can be reduced to about 2 ppm simply by storing it for about 30 min at a temperature just below its boiling-point. For water of this type, and also feed water where other methods have been inadequate in reducing the oxygen concentration to the less than 0·005–0·007 ppm required for most high pressure boilers, hydrazine treatment is the answer.

Hydrazine reacts with oxygen as follows:

$$N_2H_4 + O_2 \rightarrow N_2 + 2H_2O$$

The nitrogen liberated is not only harmless but also serves the useful function of driving oxygen out of solution, so that it can be bled off.

Some of the hydrazine tends to decompose under the action of heat as follows:

$$3N_2H_4 \rightarrow 4NH_3 + N_2$$

For this reason it is normally necessary to use double the stoicheiometric quantity of hydrazine to eliminate oxygen. As the molecular weight of hydrazine is identical to that of oxygen, namely 32, this means that two parts by weight of hydrazine are required to eliminate one part by weight of oxygen.

Hydrazine is far more expensive than sodium sulphite (see below) and for this reason one should never use more of the material than is strictly necessary.

The hydrazine–oxygen reaction is accelerated by the presence of activated carbon, and because of this filters of this material are often installed between the point where hydrazine is added, and the boiler feed water pump. Hydrazine is able to react with free ferric oxide present in the boiler plant and it is often necessary to add a considerable excess of hydrazine to the boiler when starting up.

The advantages of hydrazine are that no harmful dissolved solids are formed and that ammonia, which is liberated, is itself extremely advantageous from the point of view of reducing free CO_2 concentration.

The use of sodium sulphite
Sodium sulphite reacts readily with oxygen as follows:

$$2Na_2SO_3 + O_2 \rightarrow 2Na_2SO_4$$

In general one needs about 8 parts by weight of anhydrous sodium sulphite remove each part of oxygen or 16 parts by weight of hydrated sodium sulphite to remove each part by weight of oxygen. The reaction is catalysed by copper and cobalt salts and takes place most readily between pH 5 and 8 at elevated temperatures. Organic compounds such as glycerol or sugars tend to inhibit the reaction. Commercial sodium sulphite used for deoxygenation of boiler water usually contains 0·25% of cobalt sulphate. Sodium sulphite can be used with boilers operating up to 70 bar but it is essential in all cases to carry out good mechanical purging of dissolved oxygen to avoid excessive quantities. In general, about 50 ppm of sodium sulphite are sufficient for boilers operating at below 40 bar, whereas for boilers that operate at about 60–70 bar it is inadvisable to use more than 10 ppm of sodium sulphite, otherwise the following reaction can occur:

$$Na_2SO_3 + H_2O \rightarrow 2NaOH + SO_2$$

i.e. two extremely corrosive agents are produced. The advantages of sodium sulphite are cheapness and the fact that the sodium sulphate produced during the deoxygenation reaction is often needed to prevent caustic cracking. The disadvantage lies in the fact that quantities of sulphates are introduced, which may cause scaling in the presence of free calcium ions.

Ion exchange
There are a number of different ion-exchange resins on the market, which are able to reduce the oxygen concentration in boiler water to very low values. These are, in general, anion resins which have reduced copper or silver, ferrous hydroxide, manganous hydroxide or other reducing agents coordinated with them. As water is pumped through such resins, the oxygen concentration can be reduced to less than 0·01 ppm. Metals are regenerated

by sodium hyposulphite, and ferrous or manganous compounds with a mixture of ferrous sulphate and sodium hydroxide solutions. The method is not as yet as widely used as the other techniques mentioned above.

TREATMENT OF HIGH-PRESSURE BOILER WATER

As already mentioned, high-pressure boilers used in conjunction with steam turbines use almost total return of the condensate, losses being made up as they occur with water thoroughly demineralised by means of such processes of mixed-bed ion exchange. Oxygen is carefully removed by one of the methods mentioned, and finally the pH is raised by adding small quantities of ammonia or certain volatile amines such as cyclohexylamine. No other additions must be made. Sodium hydroxide must never be used in high-pressure boilers even in trace quantities, because of the extreme danger that it may induce caustic cracking phenomena.

The disadvantage of ammonia is that it may cause slight attack on copper fittings. Neither cyclohexylamine nor morpholine suffer from this disadvantage; however they are more expensive. The general aim is to keep the pH of the boiler water at about 9. If carbon dioxide is absent, this can be achieved by the following additions:

$$0\cdot2 \text{ ppm of ammonia} \qquad \text{or}$$
$$1\cdot0 \text{ ppm of cyclohexylamine} \qquad \text{or}$$
$$4\cdot0 \text{ ppm of morpholine}$$

If carbon dioxide is present the additional quantities required per unit weight of carbon dioxide are $0\cdot7$ units by weight of ammonia or 2 units by weight of cyclohexylamine or morpholine.

THE USE OF OCTADECYLAMINE

Octadecylamine, which has the chemical formula $C_{18}H_{37}NH_2$ is able to reduce the amount of corrosion in boilers by over 90%, and can be used with boiler temperatures in excess of 500°C. Octadecylamine however, has the unfortunate property of detaching rust deposits already formed, which may at times be harmful as the loose rust may block up traps, etc. The feed rate usually employed with regard to octadecylamine in the boiler water is maintained at about 1–2 ppm, at which concentration the required corrosion-resistant surface film on metals is not unduly disturbed.

MEDIUM- AND LOW-PRESSURE BOILERS

For boilers operating at about 300°C and below, either cyclohexylamine or morpholine is preferred as an additive. For CO_2-free boiler water either 1 ppm of cyclohexylamine or 4 ppm of morpholine will achieve the desired pH of 9. These substances have advantages over ammonia in that they do not attack copper fittings and are not as volatile as ammonia. Their cost is, however, much higher.

CAUSTIC CRACKING

This is a phenomenon which can affect a boiler plant, where the water contains free sodium hydroxide, even if this is present only in minute proportions. The sodium hydroxide orginates often from sodium carbonate, added for water-treatment purposes:

$$Na_2CO_3 + H_2O \rightarrow 2NaOH + CO_2$$

a reaction which takes place at elevated temperatures. Any minute cracks or grain boundaries in a stressed section of a boiler may be affected. The exceedingly dilute caustic solution penetrates such cracks and the water evaporates, leaving the caustic soda behind. This process is repeated until the concentration of the solution inside a crack may reach about 5–10%. Caustic cracking can take place with all plain carbon steels containing up to 0·5% carbon and also with alloy steels and stainless steels, with the exception of the 12–14% chromium types. Nickel can also be affected, but other constructional materials including cast iron appear to be immune. The reaction that takes place is an electrochemical one in which the cell can be represented by:

iron/conc. NaOH soln/dil. NaOH soln/iron

The anode is the iron in contact with the concentrated NaOH solution, i.e. the crack. Iron is dissolved by 10% sodium hydroxide to form sodium ferroate as follows:

$$2NaOH + Fe \rightarrow Na_2FeO_2 + H_2$$

The net effect of caustic cracking is that the steel structure is made progressively weaker, until a sudden surge of pressure inside the boiler produces an often disastrous boiler explosion. The main methods of preventing caustic cracking are the following.

The addition of sodium sulphate
This is added to the boiler water in a sufficient quantity to ensure that the weight ratio, $Na_2SO_4/NaOH$, always exceeds 2·5. The sodium sulphate is frequently added as such. Alternatively either sulphuric acid or magnesium sulphate—are added to boiler water already containing free NaOH. The reactions are the following:

$$2NaOH + H_2SO_4 \rightarrow Na_2SO_4 + 2H_2O$$

and

$$2NaOH + MgSO_4 \rightarrow Na_2SO_4 + Mg(OH)_2$$

It is vitally important that the water be analysed accurately so that the amounts of chemicals added as not in excess of the quantities needed.

Sodium sulphite is often added to such water to reduce the oxygen content and this also produces sodium sulphate according to the following equation:

$$2Na_2SO_3 + O_2 \rightarrow 2Na_2SO_4$$

The mechanism by which sodium sulphate prevents caustic cracking is that, instead of sodium hydroxide alone penetrating the crack, it is now a mixture of sodium sulphate and sodium hydroxide which enters. Long before the sodium hydroxide solution reaches its critical concentration of about 10%, the sodium sulphate crystallises thus barring the entry of any further quantities of solution into the crack or intergranular space. In this way it becomes impossible for the solution to reach its critical concentration of 10% NaOH.

The disadvantage of sodium sulphate treatment is that calcium sulphate may be formed which is one of the main scaling agents of boiler plant.

The addition of sodium phosphates
Polyphosphate $(NaPO_3)_6$ is added to boiler water and can react as follows:

$$(NaPO_3)_6 + 6H_2O \rightarrow 6NaH_2PO_4$$

The sodium dihydrogen phosphate formed then neutralises the calcium or sodium hydroxide introduced in the water softening process. Free calcium ions in solution form sparingly soluble calcium phosphate which is precipitated without scale formation. Polyphosphate treatment is carried out in such a way that there is always an excess of about 10% of NaH_2PO_4 in solution in excess of any hydroxide present. It is, however, important not to overdose otherwise the pH of the boiler water may fall below the recommended level.

The addition of sodium nitrate
For boilers operating at pressures between 7 and 40 bar, sodium nitrate is added to the boiling water so that the weight ratio, $NaNO_3/NaOH$, is maintained above unity. Sodium nitrate appears to inhibit caustic cracking extremely effectively, as it makes portions liable to be affected passive.

The addition of organic agents
For low pressure boilers caustic cracking can be avoided by the addition of certain organic reagents such as tannin, lignin, quebracho, etc. The function of these is, like that of sodium sulphate, to enter cracks and intergranular apertures, thereby preventing the harmful build-up of sodium hydroxide. The method is, however, only of use in boilers operating at a pressure below about 20 bar.

The use of crack-resisting steels
Certain steels, such as those which have had a reasonable quantity of aluminium added during manufacture to eliminate traces of intergranular iron oxide formation, appear to be resistant to caustic cracking. There are a number of such steels on the market and they should be used for the construction of sections in boilers, which are particularly subject to caustic cracking phenomena.

o

10.2 Various practical methods of preventing corrosion in boilers

LOW-PRESSURE BOILERS

Corrosion in low-pressure boilers is prevented by maintaining strong alkalinity, in that the sum of the sodium hydroxide and sodium carbonate contents is equal to some 12% of the total weight of solids dissolved in the water. Sodium sulphate has, of course, to be present to avoid caustic cracking. When deaeration is practiced, so that the oxygen content is below 0·05%, the sum of the weights of sodium carbonate, sodium hydroxide and trisodium phosphate should be about 10% of the total solid contents.

COPPER IN FEED WATER

Feed water to power stations often contains up to 0·2 ppm of copper. This has been found to increase the rate of corrosion considerably. Modern ion exchange methods are able to reduce the copper concentration to about 0·002 ppm, at which figure the adverse effect of copper is much reduced.

IDLE BOILERS

When boilers are left idle, oxygen may tend to build up in the water. This can cause excessive crevice corrosion. When a boiler is laid up for a short while, about 150 ppm of Na_2SO_3 and the same amount of alkali are added as under normal working. When boilers are laid up for longer periods of time, 100 ppm of sodium hydroxide and 250 ppm of sodium sulphite are added. All valves which may admit air are closed and the sulphite concentration should be tested weekly to see that its concentration does not fall below 50 ppm for short period lay-ups and 100 ppm when the plant is laid up for longer.

An alternative method is the addition of 0·1% of sodium nitrite to an alkaline solution of boiler water. Shell boilers can be kept for several years in this way, without marked deterioration occurring. The solution must, however, be discarded when the boiler is recommissioned.

CONDENSATE CORROSION

When steam condenses at the end of its working cycle, it may pick up a number of gases that cause corrosion in condensers.

The gases normally found in condenser water are: ammonia, carbon dioxide, oxygen, hydrogen, hydrogen sulphide and sulphur dioxide. Of these oxygen tends to be present only in very small quantities in high-pressure boilers where effective methods have been used to exclude it from the feed water. Even in medium- and low-pressure boilers the concentration of oxygen is normally quite low. However, even trace quantities of dissolved oxygen in condensate water have very harmful effects.

Carbon dioxide is formed by the decomposition of carbonates and bi-carbonates in feed water. It makes the condensate acidic, and this causes rapid attack on such stressed sections as the threaded joints of pipelines. The harmful effect of carbon dioxide in condensate lines can be prevented by adding neutralising amines such as cyclohexylamine, morpholine or octadecylamine. These are added in sufficient quantities to keep the pH above 7.

Ammonia may be present in the condensate due to decomposition of hydrazine, added for the purpose of deoxygenation, or nitrogenous matter in the feed water. If there is a very high concentration of ammonia, copper fittings may be attacked. Normally, however, ammonia is beneficial as it is able to react with free carbon dioxide.

The effects of sulphur dioxide and hydrogen sulphide are the same as those of carbon dioxide. They cause the condensate water to become acid, thus increasing the rate of corrosion particularly at stressed sections. As for carbon dioxide, their harmful effects can be countered by the addition of neutralising amines.

Hydrogen is produced by the direct reaction between steel and hot water and is quite harmless.

10.3 Corrosion on the fire side of a boiler

This section was written by Mr. A. E. Lock, MInstF, FIPlantE, Chief Engineer, Combustion Chemicals Limited, Chertsey, Surrey, England.

SOURCES OF CORROSION

Coal

Coal contains sulphur, chlorine, phosphorus, fluorine, arsenic and carbonates. The chlorine content may vary from traces up to about 1·2% and is normally in the form of NaCl but may also be present as KCl, $CaCl_2$ or $MgCl_2$. The use of coals containing less than 0·3% of chlorine is not liable to lead to the formation of appreciable boiler deposits, but with a higher chlorine content trouble is to be expected particularly with high-moisture-content coals. Sodium chloride, being water soluble, is washed from the coal and can settle and dry out in any part of the handling system. In chain-grate stokers fouling and jamming of links can occur. Some chlorides present in coal decompose on heating to form HCl gas. Sodium and potassium also vaporise at flame temperatures and can condense on the tube surfaces where they are converted to the appropriate sulphates and acid sulphates by any SO_2 and SO_3 present.

The average sulphur content of British coal is 1·6%. Hydrogen present in coal is converted into water vapour during combustion. This creates a humid condition affecting the dew-point of the combustion gases, so that a reaction takes place with the sulphur dioxide emitted and free oxygen, to form sulphuric acid. The latter readily attacks metal surfaces when they are below the dew-point of the combustion gases, i.e. when starting up from cold, or when fires are banked. As the boiler heats up, the iron sulphate formed crystallises as scale on the heating surfaces. It then interferes with heat transfer and reduces the area of the gas passages. The corrosive action of hydrochloric acid is also markedly increased when sulphuric acid is present.

Phosphorus is present in coal in quantities varying from 0·01 to 0·15%. Phosphates volatise at temperatures above 1,600°C and assist in the formation of bonded deposits on boiler tubes. Fluorine is usually present in quantities of up to 150 ppm and contributes to this.

Finally, ash itself contributes to corrosion problems.

Oil

Heavy fuel oils may contain more than 3·5% of sulphur, which burns to form oxides which may react with the ash and some hydrocarbons on the tube surfaces of the boiler. Vanadium and sodium are also present in fuel oils and the ash fusion points of the vanadium and alkali sulphates are often below the temperature of the metals with which they come into contact. In consequence, a slag is formed against the metal surfaces. This may take place at about 480–540°C. At these temperatures vanadium pentoxide acts as a flux to lower the ash fusion temperature so that a hard coating is formed over the metal surfaces. This causes high-temperature corrosion, which normally takes place on superheater tubes, at the beginning of the first pass of tube boilers, or anywhere else where metal temperatures are high.

LOW-TEMPERATURE CORROSION

In modern boiler plants the temperature of the flue gases is usually reduced either before entering the chimney, or inside the chimney, to below the dew-point of the flue gases. This means that free H_2SO_3 and free H_2SO_4 are formed. The former is only produced at very low temperatures, but sulphuric acid is formed extremely readily. It can seep through deposits and attack metal tubes as follows:

$$Fe + H_2SO_4 \rightarrow FeSO_4 + H_2$$

causing rapid attack. Some of the sulphuric acid reacts with metal oxides present in the fuel to form hygroscopic metal sulphates, which again induce more rapid corrosion.

Normally the dew-point of sulphuric acid in the flue gases is about 130–160°C. Sulphur dioxide, which is always present in the flue gases, is oxidised under the catalytic action of the furnace deposits, which contain appreciable quantities of V_2O_5, to SO_3 and hence into sulphuric acid. Even the corrosion products act as catalysts, changing more sulphur dioxide into sulphuric acid.

It is possible to reduce the formation of sulphur trioxide, and thus sulphuric acid by reducing the excess of air for combustion down to about 0·5–1·0% oxygen. A large boiler plant, which is running at a steady load, may be operated with almost stoicheiometric proportions of free oxygen, so that the formation of sulphuric acid in the furnace is very much reduced. It is, however, virtually impossible to run plants efficiently where the load fluctuates. Even with only 0·2% of oxygen in excess, tests have shown the presence of 6–12 ppm of SO_3 in the flue gases, depending on the amount of sulphur present in the fuel.

Corrosion of furnaces can be prevented by two general methods:

1 The purification of the fuel used, which involves the removal of sulphur.
2 The addition of chemicals to the combustion chamber or to the fuel used.

Purification of the fuel used
The methods of removing sulphur from heavy fuel oil are as follows.

1 Hydro-desulphurisation in which hydrogen is used together with cobalt molybdenite as a catalyst to convert any sulphur present in the oil into hydrogen sulphide according to the following equation:

$$H_2 + S \rightarrow H_2S$$

The hydrogen sulphide is taken off and the sulphur utilised for sulphuric acid manufacture. It was stated in 1968 that the average cost of desulphurisation was equal to 75p ($1·80) per ton of fuel used for every 1% reduction in sulphur content. A plant capable of treating one million tons of fuel oil per annum costs about £4·5 million. ($10·8 million).

2 The Cat-Ox system, developed by Monsanto of St Louis, Missouri, USA, utilises a converter plant where any sulphur present is oxidised to sulphur dioxide, which is then further oxidised to sulphur trioxide and absorbed in concentrated sulphuric acid to form oleum. The plant is, in actual fact, a miniature sulphuric acid contact plant. No details of cost are given.

Under present economic conditions it seems unlikely that sulphur removal from the fuel oil fraction will be practised.

Chemical additives
Dolomite—which is composed of approximately 45% magnesium carbonate and 54% calcium carbonate has been used with considerable success, as it reacts at its surface with free sulphuric acid to give high fusion point sulphates with light friable deposits instead of heavy bonded and corrosive scale. Pure magnesia or a higher proportion of magnesium carbonate would give better results, but the chemicals are more expensive. In the case of one test, when coal with a sulphur content of 2·2% was used, 27 ppm of SO_3 were present in the exit gases. This concentration was reduced to 8 ppm by the use of dolomite. An alternative consideration shows that

this corresponds to a conversion of SO_2 to SO_3 of 0·6% as against 3% in the case of untreated fuel.

Alumag CH 22—is a suspension of finely divided particles of magnesia and alumina, and modifies the structure of the molten sodium vanadyl vanadate/sodium sulphate slag to produce an ash which has a fusion temperature in excess of 1,550°C. This porous, flaky and friable ash with a cubic crystal structure is easily removed by normal soot blowing. In addition, the reaction prevents the formation of strongly cohesive deposits. It has been successful in alleviating superheater problems and suppressing SO_3 and acidic smut formation.

Primary amines—are added to the oil in order to prevent the oxidation of SO_2 to SO_3. The primary amines dissociate during combustion and the nitrogen reacts preferentially with atomic oxygen present, thereby restricting the formation of SO_3, and consequently the sulphuric acid produced. A trade name for such an additive is 'CCL concentrate'. The addition of this material has been found to reduce the SO_3 content in the combustion chamber by half, and at the economiser to about one quarter of the value obtained with untreated fuel.

Chemicals—can be added in the combustion chamber or to the fuel oil to reduce the ignition point of carbon. If, at the same time, the SO_3 content is lowered as well, soot deposition and acidic smuts are much reduced. This is done by adding methyl cyclopentadienyl manganese tricarbonyl, which was developed by the Ethyl Corporation of America, and is sold under the trade name 'Tri-ad'. The effect of adding the material to fuel oil is to accelerate the conversion of carbon to CO_2. The metal oxides produced during the burning of Tri-ad act as surface catalysts enabling the hydrocarbons in the oil to burn out more completely. Apart from halving the amount of SO_3 produced, the addition of Tri-ad also raises the overall combustion efficiency of a furnace by up to 2·75%.

Literature sources and suggested further reading

1   *Corrosion and deposits in coal and oil fired boilers*, Battelle Memorial Institute, Ohio (1959)
2   JOHNSON, H. R., and LITTLE, D. J., *Mechanism of corrosion by fuel impurities*, Butterworths, London (1963)
3   WILLIAMS, J. N., *Steam generation*, Allen & Unwin, London (1969)
4   HAMER, P., JACKSON, J., and THURSTON, E. F., *Industrial water treatment practice*, Butterworths, London (1961)
5   FRANCIS, W., *Boilerhouse and power-station chemistry*, Arnold, London (1962)

Index

193